Wissenschaftsethik und Technikfolgenbeurteilung
Band 22

Schriftenreihe der Europäischen Akademie zur Erforschung
von Folgen wissenschaftlich-technischer Entwicklungen
Bad Neuenahr-Ahrweiler GmbH
herausgegeben von Carl Friedrich Gethmann

M. Decker · M. Ladikas (eds)

Bridges between Science, Society and Policy

Technology Assessment – Methods and Impacts

 Springer

Editor of the series

Professor Dr. Dr. h.c. Carl Friedrich Gethmann
Europäische Akademie GmbH
Wilhelmstraße 56, 53474 Bad Neuenahr-Ahrweiler, Germany

Editors

Dr. Michael Decker
Forschungszentrum Karlsruhe, Institut für Technikfolgenabschätzung und
Systemanalyse (Institute for Technology Assessment and System Analysis)
P.O. box 36 40, 76021 Karlsruhe, Germany

Dr. Miltos Ladikas
Europäische Akademie GmbH
Wilhelmstraße 56, 53474 Bad Neuenahr-Ahrweiler, Germany

Editing

Susanne Stephan
Europäische Akademie GmbH
Wilhelmstraße 56, 53474 Bad Neuenahr-Ahrweiler, Germany

Friederike Wütscher
Europäische Akademie GmbH
Wilhelmstraße 56, 53474 Bad Neuenahr-Ahrweiler, Germany

ISBN 978-3-642-05960-5   ISBN 978-3-662-06171-8 (eBook)
DOI 10.1007/978-3-662-06171-8

Bibliographic information published by Die Deutsche Bibliothek
Die Deutsche Bibliohek lists this publication in the Deutsche Nationalbibliografie; detailed bibliographic data is available in the Internet at <http://dnb.ddb.de>.

Typesetting: Köllen Druck+Verlag GmbH, Bonn + Berlin
Coverdesign: deblik, Berlin

Printed on acid-free paper   62/3020hu – 5 4 3 2 1 0 –

Europäische Akademie

zur Erforschung von Folgen wissenschaftlich-technischer Entwicklungen
Bad Neuenahr-Ahrweiler GmbH

The Europäische Akademie

The *Europäische Akademie zur Erforschung von Folgen wissenschaftlich-technischer Entwicklungen GmbH* is concerned with the scientific study of consequences of scientific and technological advance for the individual and social life and for the natural environment. The Europäische Akademie intends to contribute to a rational way of society of dealing with the consequences of scientific and technological developments. This aim is mainly realised in the development of recommendations for options to act, from the point of view of long-term societal acceptance. The work of the Europäische Akademie mostly takes place in temporary interdisciplinary project groups, whose members are recognised scientists from European universities. Overarching issues, e.g. from the fields of Technology Assessment or Ethic of Science, are dealt with by the staff of the Europäische Akademie.

The Series

The series "Wissenschaftsethik und Technikfolgenbeurteilung" (Ethics of Science and Technology Assessment) serves to publish the results of the work of the Europäische Akademie. It is published by the academy's director. Besides the final results of the project groups the series includes volumes on general questions of ethics of science and technology assessment as well as other monographic studies.

Acknowledgement

The project TAMI (Technology Assessment in Europe; between Method and Impact) has been funded by the European Commission, DG-Research, within the programme "Improving the Human Potential; Strategic Analysis of Specific Political Issues". We would like to thank the policy makers (both national and European) that took part in various TAMI discussions, particularly Ms Eryl McNally, Mr Gerhard Schmid, Mr Josef Bugl, Mr Otto Bode, Mr Leo Bjoernskov, Mr Paul Berckmans and Mr Anders Moller. We would also like to thank our scientific officer, Mr Belmiro Martins, for his support in this activity.

The project TAMI comprised the following European institutes:

Europäische Akademie GmbH (EA), Germany
Miltos Ladikas, Susanne Stephan
(susanne.stephan@dlr.de)

Parliamentary Office of Science and Technology (POST), UK
David Cope (COPED@parliament.uk)

Institute for Technology Assessment and System Analysis (ITAS), Germany
Armin Grunwald (Armin.Grunwald@itas.fzk.de)

Centre for Technology Assessment at the Swiss Science and Technology Council (TA-SWISS), Switzerland
Sergio Bellucci, Danielle Bütschi (TA@swtr.admin.ch)

Center of Technology Assessment in Baden-Württemberg (CTA), Germany
Rainer Carius (rainer.carius@ta-akademie.de)

Danish Board of Technology (DBT), Denmark
Søren Gram, Lars Klüver (sg@Tekno.dk)

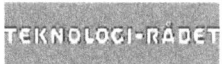

Office of Technology Assessment at the German Parliament (TAB), Germany
Leonhard Hennen (hennen@tab.fzk.de)

Committee on Industry, External Trade, Research and Energy, European Parliament (EP), Belgium
Theodoros Karapiperis (tkarapiperis@europarl.eu.int)

Centre of Science, Technology, Society Studies at the Institute of Philosophy, Academy of Sciences of the Czech Republic (STS Centre), Czech Republic
Petr Machleidt (stsscz@cesnet.cz)

Warsaw School of Economics – Institute of Modern Civilisation (SGH), Poland
Tomasz Szapiro (tszapiro@sgh.waw.pl)

Consejo Superior de Investigaciones Científicas (CSIC)
Laura Cruz-Castro, Luis Sanz-Menéndez (Laura.Cruz@iesam.csic.es)

Rathenau Institute, The Netherlands
Jan Staman, Rinie van Est (j.staman@rathenau.nl)

Flemish Institute for Science and Technology Assessment (viWTA), Belgium
Robby Berloznik, Stef Steyaert (viWTA@vlaamsparlement.be)

For more information, please contact the project manager:
Europäische Akademie GmbH
Susanne Stephan
Phone: +49/2641 973 323
Fax: +49/2641 973 320
Email: susanne.stephan@dlr.de
Wilhelmstr. 56
53474 Bad Neuenahr-Ahrweiler
Germany

Foreword

The Europäische Akademie zur Erforschung von Folgen wissenschaftlich-technischer Entwicklungen Bad Neuenahr-Ahrweiler GmbH (european academy) is concerned with the scientific study of the consequences of scientific and technological advance for the individual and social life and for the natural environment. The main focus is to examine foreseeable mid- and long-term processes that are especially influenced by natural- and engineering sciences and the medical disciplines. The academy fulfills this task by organizing interdisciplinary expert discussions.

Another important issue of the work of the Europäische Akademie concerns the methodology of Technology Assessment as a general issue. This is the main reason that the european academy organized during the past two years a project funded by the European Commission on *Technology Assessment. Methods and Impact* (TAMI). Together with partners from all over Europe a common understanding of what Technology Assessment (TA) is supposed to do was developed. Most importantly it was acknowledged that the core of any TA activity has to be a sound scientific understanding of the relevant phenomena. Communication then is of cordial importance to reach the relevant decision makers as well as the general public. It is true that this phase of the TA process has been treated with too little attention for many years. The communication processes between scientific advisers and policy makers have hence to be further scrutinized.

The book series on ethics of science and technology assessment, "Wissenschaftsethik und Technikfolgenbeurteilung", edited by the Europäische Akademie, gives the opportunity to publish the results of the TAMI project as a contribution to the ongoing debate on the methodology in European Technology Assessment.

Bad Neuenahr-Ahrweiler, January 2004   Carl Friedrich Gethmann

Preface

Today, technological impact assessment is more necessary than ever before. Technological leaps are now no longer measured in centuries, but in decades. Assessing such developments is becoming ever more complicated and, for political authorities, ever more difficult. There is no doubt that science and technology have become more intrusive and more a part of our daily lives and no-one is unaware of the power that discoveries in science have to radically alter our lives and our relationship with our environment. Information technology, biotechnology and genetic engineering are cases in point; but technological developments in agriculture, the energy sector and transport, for example, are also becoming increasingly complex.

Politicians are called upon increasingly to make decisions on matters of considerable technological complexity, and more and more on matters which involve very difficult ethical and environmental considerations. When the political community takes a decision to accept and encourage a technology, this is often not a decision which can be reversed at short notice. If Members of Parliament are to be able to take decisions on these matters in a fully informed and meaningful way, they need access not only to information, but to information which is presented in a way that can be understood by the non-expert and in a way which allows the consideration of various political options in the light of the possible consequences of those options.

Technological impact assessment requires long-term comprehensive investigation, which cannot be carried out either by a documentation centre or by research support units of a parliament. Technology Assessment performs a different role in a parliamentary context than do the standard parliamentary research services – many parliaments in Europe and elsewhere recognise this by having a separate Technological Impact Assessment Unit of some kind. Many people, including politicians, do not understand this difference in role and it is part of our duty to make the difference clear between the simple provision of information – which is certainly not a role to be denigrated or downgraded – and the assessment of scientific and technological advances, which goes a step further.

As the power of science transforms our lives in ways which humanity never before imagined, we the politicians together with the scientists carry increasingly heavy responsibility for guiding and illuminating the way in which our society develops and grows. The objective of TAMI to create and promote a structured dialogue, an interaction and information sharing within the Technology Assessment community on a more international level, is the only possible way to meet the future needs. I salute the successful completion of the project and I look forward to disseminating these results in the policymaking community of Europe. TAMI has been another step in the improvement of policy advice and therefore, better science policy in Europe.

Brussels, November 2003   Gerhard Schmid, Vice-President European Parliament

Table of Contents

Technology Assessment in Europe; between Method and Impact – The TAMI Project

Michael Decker and Miltos Ladikas

Introduction

Technology Assessment (TA) is usually described as problem oriented research. This is not however a proper portrayal of TA since many research endeavours can be described as problem oriented. When, for instance, a natural scientist designs an experiment in order to verify a theory or an engineer plans the construction of a bridge, this can also be described as problem oriented research. While these examples represent scientific or technical problems, TA has to deal also with social, political or environmental problems, which are embedded outside science and technology but derive from and refer to scientific issues. Moreover, TA does research on the potential consequences of these developments, the use and the disposal of new technical artefacts or on the development, use and output of new technical production systems. TA contributes to the decision but not necessarily to the technical realisation. In the example of planning the construction of a bridge, TA would deal with the basic question "is there a need for a bridge crossing to the river?" and consider this against the alternative options (e.g. ferry). If necessary, TA would delve into technical aspects (e.g. by comparing the pros and cons of a suspension, rotating or draw bridge). The argumentative ground for these decisions refers to consequences (health, social, economic, environmental, etc.) resulting from the different options.

At the dawn of official TA thirty years ago, TA was focusing on concrete predictions of technological consequences. The aim was to gain advance knowledge on technology options in order to make better (i.e. better informed) decisions. The "early warning" aspect was central to TA business in order to steer clear of identified potential hazards or at least to minimize their effects[1]. The Office of Technology Assessment (OTA), specifically established to provide scientific advice to the US Congress, represents the "classical" TA approach that contributes to the political decision making process by providing comprehensive knowledge on the consequences. This is illustrated by the reasoning for the creation of the institute: "it is essential that, to the fullest extent possible, the consequences of technological applications be anticipated, understood, and considered in determination of public policy on existing and emerging problems" (United States Senate 1972). OTA's initial functions are still valid nowadays within the TA scene and include the identification of impacts of technology, assertion of cause-and-effect relationships and identification of alternative programs and options. In this manner, the division of

---

[1] Harremoes et al. (2002) sampled very impressive case studies from the past hundred years where early warnings have been missing or failed.

labour between TA and decision making is clear, whereby TA "… gives no recommendations concerning what should be done but […] provides information what could be done" (Gibbons 1991, p 27).

Overall, the beginning of TA as scientific advice can be described as the effort to develop a comprehensive set of "laws of progress" so that the state can direct technical development (Schmid et al. 2003). In the last two decades, European TA developed several methodological concepts that aim, in general, to enhance the initial approach of "classical TA". Grunwald (1999) has attempted to classify the variety of methodologies into:

The system analysis approach; a multi-step method which at the beginning of European TA was taken as the main paradigm for TA (Jochem 1975; Paschen et al. 1978) and aimed "…as most important research area of system analysis at analysing the potentials and consequences of technical developments" (Eberlein 1995, p 11).

TA as strategic framing concept (Paschen et al. 1978; Paschen et al. 1987; Paschen and Petermann 1991) that takes into account scenario techniques in order to accompany the development of new technologies.

The Technology Assessment of the German Association of Engineers (VDI) that developed an octagon of values referring to feasibility of technical systems, economic efficiency of technical decisions, general economic prosperity, safety and survival capabilities of individual users and the whole mankind, health (including mental and physical well-being), environmental quality aspects, and the possibilities of development of individual personalities and the society as whole (VDI 1991).

Participatory TA methods, which enhanced the scientific oriented TA process by incorporating stakeholders and laypersons. With these methods, TA turned from the primary cognitive questioning (related to a kind of optimizing for decisions by development of relevant criteria) to an endeavour which aims at or negotiates for reaching consensus or compromises in conflict situations concerning technology (Baron 1995).

As mentioned above TA is problem oriented research which aims to contribute to solutions of political, social, ecological etc. problems, i.e. problems deriving from outside the realm of science and technology. The results of TA are, in general, concrete recommendations to policy makers and therefore the original questioning as well as the area of potential impact is outside the scientific system. The shortcomings of the "classical" TA approach can be summarised in the fact that the whole TA-process (starting from the "transformation" of the extra-scientific problem into a scientifically manageable research programme until the feedback of the recommendations into the policy making process) needs relevance decisions, evaluations, and the development of criteria, which *is at least partially normative and value loaded* (Decker and Grunwald 2001). Thus, the division between value neutral scientific advice and political decision, which takes into account norms and values, cannot be kept up. As such, in the initial planning phase the optimism and hope to gain knowledge about "laws of progress" could not be fulfilled. The development of participatory TA can be described as the effort for TA to have a greater impact by handing over decisions based on values to society itself. Therefore practitioners of participatory TA could claim that their results find acceptance in science as well as stakeholders and laypeople[2]. In most participatory TA-approaches this is realized by a methodological combination of scientific and participatory discourses that influence each other in a positive way (Gottschalk and Elstner 1997).

The advance of participatory TA particularly in Denmark, The Netherlands and Switzerland has been impressive but not without its critics (e.g. Grunwald 2001; Gethmann 2001). Firstly, there are doubts whether it is possible to combine the scientific and participatory components of the TA process without weakening the scientific part of it. The main argument here is that an intense, quality controlled, interdisciplinary, scientific collaboration is of crucial importance to develop concrete recommendations to solve the problem under consideration. It is doubtful whether the scientific part of participatory TA approaches can achieve this (Decker and Neumann-Held 2003)[3]. Moreover, the issue of representation has also been raised questioning the capability of the few stakeholders/citizens participating in the discourse to represent all those that have not been part of the discourse (Grunwald 1999; Liakopoulos 2001).

More recently, the communication aspects of TA have been the focus of discussions in the TA community. These, refer to methodological developments that provide TA with the means to directly intervene in the process of opinion forming by providing new processes of societal communication and not by inducing a particular opinion, perspective or political position (van Est et al. 2002). This new debate as well as the older debates about the merits and disadvantages of classical vs. participatory TA, has been the focus of discussions within the project TAMI.

The TAMI project

The project TAMI (Technology Assessment in Europe; between Method and Impact) was created in order to answer the multitude of questions arising from the developments in TA as discussed above. On one hand, the variety of methodologies within the spectrum classical-participatory TA have been tried and have evolved throughout the years changing the face of TA irrevocably and making an overall review of their potential necessary. On the other hand, a lot of questions about core benefits of TA have been left unanswered for too long. Since TA identifies itself as policy advice and is therefore in most cases publicly funded, this already raises the question of whether TA is worth this funding. Is TA improving policy making on technological issues? How can the impact of TA in policy advice be assessed, and, according to what criteria?

TAMI attempted to explore all major issues of TA in Europe by creating a structured dialogue within the TA community as well as between TA experts and policy

---

[2] The concrete composition of the group of people participating is a crucial aspect of the respective participatory TA concept. The criteria here are for example "being concerned", "representative for the respective society", "randomized", etc.

[3] Hanekamp (2001) has made an interesting suggestion to distinguish between three different types of participation: Democratic, administrative, and research participation where the "usual" debate concerning the involvement of laypeople would then be part of research participation only.

makers. By involving an impressive array of most major TA institutes in Europe[4], the TAMI group set out to:

- review and evaluate the state-of-the-art methodologies and practices used in current TA;
- review the issue of impact and develop criteria for its evaluation;
- attempt a systematic and strategic view of the relationship between method and impact;
- identify all relevant factors that influence the functions of TA such as the institutional and political context;
- draw conclusions on a common TA "reference system" that includes the dimensions of Method, Impact and Policy.

To reach its objectives, TAMI followed the so called "twin group principle" whereby two groups, the "method group" and the "impact group", started from different perspectives without coupling. In the second half of the project the combination of the findings of the two groups was initiated in a series of "cross-over" discussions. The advantage of this procedure was that the two groups could function as evaluators for each other's results and within the common discussions it was possible to reach a certain level of commonality between the two working groups without losing the benefit of tackling the issues from two perspectives.

The method group

The method group comprised experts from most TAMI partners with particular interest in the functions and parameters of various methodologies used in European TA. The consortium of partners has been heterogeneous enough and the contextual experiences of each institute different enough to focus the discussion on basic questions about TA. The common understanding of what TA is, was the starting point of the discussions. TA has experienced a series of metamorphoses since its conception days with new aims and functions that would seem unrecognisable as TA by the early practitioners. This has led to confusion as to what exactly constitutes a TA activity nowadays and the method group had to find a common definition to start from.

The next major question has been: "how can TA-practitioners optimise their TA-projects in order to reach the impact they strive for?". Usually, a "customer" orders a TA-project on a particular topic as the basis for a concrete political decision but without any "advance commitment" to the projects results. This means that the project has to develop its own "legitimisation power", in the sense of creating

---

[4] Members of TAMI were: Europäische Akademie GmbH, Germany; Parliamentary Office of Science and Technology, UK; The Forschungzentrum Karlsruhe GmbH, Institute for Technology Assessment and System Analysis, Germany; Center of Technology Assessment in Baden-Württemberg, Germany; Danish Board of Technology, Denmark; Centre for Technology Assessment at the Swiss Science and Technology Council, Switzerland; Centre of Science, Technology and Society Studies at the Institute of Philosophy of the Academy of Sciences, Czech Republic; Warsaw School of Economics, Institute of Modern Civilisation, Poland; Consejo Superior de Investigaciones Cientificas, Spanish Policy Research on Innovation & Technology, Training and Education, Spain; Committee on Industry, External Trade, Research and Energy, European Parliament; Rathenau Institute, The Netherlands; and, Flemish Institute for Science and Technology Assessment, Flemish Parliament, Belgium.

results that are hard to ignore. In practice, one combines several TA-methods in order to realize the TA-project and takes care that the project is done in an optimal way. But, what constitutes an optimal way depends heavily on contextual parameters that the TA practitioner has to take into consideration. The method group has worked to provide answers as to the optimisation of TA projects with regard to improving their internal consistency and their eventual impact.

The impact group

What constitutes the "impact of TA" is an often discussed but scarcely investigated subject. Impact assessment exercises are rare and refer to particular projects with specific contextual variables (e.g. Oppermann 2001; Renn et al. 1999). The EUROPTA report, focusing on participatory methods (EUROPTA 2000) has identified the problem of impact assessment which is essential in proving the "worth" of participatory processes. As the first step towards evaluating impact, the authors describe an inventory of "political roles" of participatory TA, different influencing factors, point out to the decisive importance of project management (Klüver 2000) as well as the choice of the "right" participatory TA-methods (Eijndhoven and Est 2000). The EUROPTA-project can be seen as the first concerted effort to discuss impact assessment and which TAMI has heavily drawn upon its results.

The TAMI impact group comprised experts from different European TA institutes and from various contexts that had to be taken into account. Whether the institute works in direct contact with the parliament and receives concrete requests to work on, or not; whether its aim includes promotion of participation of stakeholder and citizens in the debate, or not; whether it focuses on interdisciplinary scientific assessment, or not; these are TA "trends" that demand different views of what constitutes impact. Questions such as "what is a common understanding of TA impact?", "what are the prerequisites and influencing factors involved?", "what are the universal functions or roles of TA from which impact should be expected", and, "which methodologies and evaluation procedures can best describe the desired impact?" have been the focus of discussions and results of the impact group deliberations.

The TAMI process

The TAMI project was organised on the "twin group" principle (see above) and strove for common authorship of all participants in the main report of the two groups. Agreement by everybody tends to create intense discussions not only on TA methods and their impact but also on the cultural, societal and political context TA is rooted in. Therefore, the work did also go one step further by requesting consensus.

Feedback loops

There have been various "feedback loops" in the process of the TAMI project. The "twin group" arrangement with occasional "cross over" meetings whereby each group had the chance to evaluate the work of the other group and get feedback for their work, was the main internal "feedback loop". Nevertheless, TAMI consists of

TA experts with big dependencies on outside parameters (clients, institutional context, etc.) which made it necessary to create another "feedback loop" based on exchange with external experts related to TA. Therefore, TAMI organised additional meetings and invited European, national and regional policy makers as well as representatives from industry to take part in the discussions and provide feedback on the work of TAMI.

The first external feedback took the form of a Kick-off meeting which aimed at broadening the discussion basis at the beginning of the project. There, external experts representing various policy-making communities as well as industry with intense R&D efforts[5] were asked to provide their opinion in issues such as:

- What are your expectations from TA-institutes?
- If you were to create such institute, what would be the reasons to do that and what types of product would you like it to produce?
- How would you improve the impact of TA? Could you define specific criteria for success?
- Can you give an example of an influential report?
- What are the main problems in S&T policy that TA can contribute to? And how?
- What do policymakers need for their decisions concerning S&T? What do they think about participation of experts, laypersons, citizens and/or stakeholders?
- Is there scope for collaboration between public and industry TA?

The results of the first external feedback were incorporated in the work of TAMI which was evaluated again during a Mid-term meeting. The external experts in this meeting were again representatives of the policymaking community as well as those with relevant communication expertise[6], and the topics for discussion included a range of issues deriving from the discussions within the two groups of TAMI:

The goals of TA. The discussion in the TAMI project about impact of TA brought up a variety of goals TA is striving to achieve as well as roles TA is expected to play. TAMI developed a typology of impact that shows a set of nine types of roles, ranging from "scientific assessment" of chances and risks to initialising "new policies". What is the feedback on the completeness of this typology.

Quality criteria of TA. High quality is a necessary condition for the viability of TA and TAMI created a list of quality criteria that need to be fulfilled. What is the trade-off between quality and speed that appears a necessary prerequisite from the clientele of TA?

From scientific to communication process. The TAMI project identified communication as a central aspect of the TA mission. Communication aspects of TA refer to

---

[5]   These were: Andrew Freeman (GlaxoSmithKline), Paraskevas Caracostas (DG Research, Foresight Unit, European Commission), Hans Peter Bernhard (Novartis Services AG) and Josef Bugl (former Member of the German Bundestag, Chair of the Advisory Board of the Centre for Technology Assessment Baden-Württemberg).

[6]   The following external evaluators participated in the discussion: Eryl McNally (Member of the European Parliament), Cees Midden (Eindhoven University of Technology), Michael Nentwich (Austrian Academy of Sciences), Paul Berckmans (Flanders' Social and Economic Council) and Otto Bode (German Federal Ministry of Education and Research).

both the process and the output of TA. As far as the output of TA is concerned, TA has an urgent need for proper communication strategy that would increase the reception of the output and therefore the overall impact of TA. Concerning the TA process, in some parts of Europe TA changes from supporting decision-making to raising awareness, stimulating public debates and expanding knowledge amongst the general public can be observed. How does the shift of focus affect the overall impact of TA?

Flexibility. TA projects start with a situational appreciation. However, this cannot be taken for granted during the whole project. Communication is necessary for keeping track both with the ongoing scientific/stakeholder debate and the political/social debate. Are TA projects flexible enough to adjust to rapid changes in the social and scientific debate as well as the policy-making process?

The results of this second external "feedback loop" were incorporated into the final outcome of the project which is presented in this volume.

About this volume

The present volume is divided into three main parts, each representing a particular outcome of the TAMI project.

Part I: Main results

This part includes the main results of the project. The struggle for common authorship has been successful and denotes the willingness of the TA community to overcome differences and reach consensus on very important matters for the functions and future of TA.

The Method Group results are presented under the title "The Practice of TA. Science, Interaction, and Communication". This chapter deals with the main issue of a common understanding and definition of TA, review of methodologies, development of guidelines for project design, quality criteria and project implementation.

The second chapter titled "Towards a Framework for Assessing the Impact of Technology Assessment" is the main outcome of the Impact Group deliberations and covers issues such as common understanding and definition of impact, a detailed typology of impacts, description of influencing factors other than method, and communication aspects.

Finally in the chapter "Conclusions, Recommendation & Wider Perspectives", there is a main summary of the results from both groups and recommendations for actions in relation to the evaluation of impact, communication functions of TA and trans-national collaboration.

Part II: Supplementary papers

This part includes papers worked out by sub-groups of the Impact Group on themes that appeared to be of particular importance in the process of TAMI and where further exploration was deemed necessary.

The first paper "Shaping the impact: the Institutional Context of Parliamentary Technology Assessment" by Laura Cruz-Castro and Luis Sanz-Menéndez explores the influence of the institutional context which parliamentary TA is required to work in. The authors describe case studies from various national contexts and develop conceptual categorisations which lead them to conclusions about the impact of TA in the parliaments and appropriate adaptation strategies.

The following paper "Organised Interests in the European Union's Science and Technology Policy; The influence of lobby activities" by Theo Karapiperis and Miltos Ladikas focuses on the issue of lobbying influence in policymaking in relation to the basic roles of TA. The authors analyse a survey of European members of the parliament and follow a series of case studies at the European Parliament that show both the conflicts and the possibilities for synergies between lobbyists and TA experts.

The paper "Industry Technology Assessment: Opportunities and Challenges for Partnership" by Robin Fears and Susanne Stephan deals with the issue of industry TA, its functions and similarities with public TA. A review of the recent debate and a case-study form the health care sector in the UK, provides the authors with arguments to suggest a road forward towards closer collaboration between the two TA traditions.

The final supplementary paper "Culturally-based Framing Factors that Influence Technology Assessment" by Tomasz Szapiro provides an overview of the literature and a paradigm of cultural analysis based on the "matrix of impacts" as developed in TAMI and adapted to the work of technology assessment.

Part III: Appendix

The annex includes a one page description of each institute participating in TAMI. Since the institutional setting has been identified as crucial momentum within the relation of method and impact this annex demonstrates the particular variety of TAMI-institutions.

It also includes a series of examples for the roles of TA described in the typology matrix of the impact group deliberations. The immense experience of TAMI-partners has been activated to find examples of real-life cases for most of the roles identified in the typology. As a service to the reader, these examples incorporate full information on references and contact details.

References

Baron W (1995) Technikfolgenabschätzung. Ansätze zur Institutionalisierung und Chancen der Partizipation. Westdeutscher Verlag, Opladen

Büllingen F (1999) Office of Technology Assessement (OTA). In: Bröchler S, Simonis G, Sundermann K (eds) Handbuch Technikfolgenabschätzung, Edition Sigma Berlin, pp 411–416

Bütschi D, Nentwich M (2000) The Role of PTA in the Policy-Making Process. In: EUROPTA-Report, Copenhagen. www.tekno.dk/europta

Casper BM (1986) Anspruch und Wirklichkeit der Technikfolgenabschätzung beim US-amerikanischen Kongress. In: Dierkes M, Petermann T, Thienen V (eds) Technik und Parlament. Technikfolgenabschätzung: Konzepte, Erfahrungen, Chancen. Edition Sigma, Berlin, pp 205–237

Decker M, Grunwald A (2001) Rational Technology Assessment as Interdisciplinary Research. In: Decker M (ed) Interdisciplinarity in Technology Assessment. Implementation and its Chances and Limits. Springer, Berlin, pp 33–60

Decker M, Neumann-Held EM (2003) Between Expert TA and Expert Dilemma. A Plea for Expertise in Technology Assessment. In: Bechmann G, Hronsky I (eds) Expertise and Its Interfaces. The Tense Relationship of Science and Politics. Edition Sigma Berlin, pp 203–223

Eijndhoven J van, Est R van (2000) The choice of participatory TA methods. In: EUROPTA-Report. www.tekno.dk/europta

Gethmann C F (2001) Participatory Technology Assessment. Some critical questions. In: Decker M (ed) Interdisciplinarity in Technology Assessment. Implementation and its Chances and Limits. Springer, Berlin, pp 3–13

Gibbons J (1991) Technology Assessment am Office for Technology Assessment: Die Entwicklungsgeschichte eines Experimentes. In: Kornwachs K (ed) Reichweite und Potential der Technikfolgenabschätzung. Stuttgart, pp 23–48

Gottschalk N, Elstner M (1997) Technik und Politik. Überlegungen zu einer innovativen Technikgestaltung. In: Elstner M (ed) Gentechnik, Ethik und Gesellschaft. Heidelberg, pp 143–180

Grunwald A (1999) Technikfolgenabschätzung. Konzeptionen und Kritik. In: Rationale Technikfolgenbeurteilung. Konzepte und methodische Grundlagen. Springer, Berlin

Hanekamp G (2001) Scientific Policy Consulting and Participation. In: Newsletter No 24, Europäische Akademie GmbH, Bad Neuenahr-Ahrweiler

Harremoes P, Gee D, MacGarvin M, Stirling A, Keys J, Wynne B, Guedes Vaz S (eds) (2002) The Precautionary Principle in the 20th century. Late Lessons from early warnings. EARTHSCAN, London

Hennen L (2000) Impacts of participatory TA on its societal environment. In: EUROPTA-Report, Copenhagen. www.tekno.dk/europta

Jochem (1975) Möglichkeiten und Grenzen der Technikfolgen-Abschätzung und -Bewertung (TA) – dargestellt an einigen ihrer Grenzen. In: Haas H (ed) Technikfolgenabschätzung Köln, pp 55–66

Klüver L (2000) Project Management – a matter of ethics and robust decisions. In: EUROPTA-Report. www.tekno.dk/europta

Liakopoulos M (2001) The Politics of Technology Assessment. In: M. Decker (ed) Interdisciplinarity in Technology Assessment; implementation and its chances and limits. Springer, Berlin

Müller A, Tulickas E, Wienhöfer E (1996) Vorläufige Bewertung des Verfahrens "Bürgerforum". In: Wienhöfer E (ed) Bürgerforen als Verfahren der Technikfolgenbewertung. Arbeitsbericht der TA-Akademie Stuttgart, pp 115–128

Paschen H, Gresser K, Conrad S (1978) Technology Assessment: Technologiefolgenabschätzung. Ziele, methodische und organisatorische Probleme, Anwendungen. Campus, Frankfurt New York

Paschen H (1986) Technology Assessment – Ein strategischen Rahmenkonzept für die Bewertung von Technologien. In: Dierkes M, Petermann T, Thienen V (eds) Technik und Parlament. Technikfolgenabschätzung: Konzepte, Erfahrungen, Chancen. Edition Sigma, Berlin, pp 21–46

Paschen H, Bechmann G, Wingert B (1987) Funktion und Leistungsfähigkeit des Technology Assessment im Rahmen der Technologiepolitik. In: Kruedener J, Schubert K von (eds) Technikfolgen als sozialer Wandel. Köln, pp 57–73

Paschen H, Petermann T (1991) Technikfolgenabschätzung – ein strategisches Rahmenkonzept für die Analyse und Bewertung von Technikfolgen. In: Petermann T (ed) Technikfolgen-Abschätzung als Technikforschung und Politikberatung. Campus, Frankfurt, pp 19–42

Renn O, Webler T (1998) Der kooperative Diskurs. Theoretische Grundlagen, Anforderungen, Möglichkeiten. In: Renn O, Kastenholz H, Schild P, Wilhelm U (eds) Abfallpolitik im kooperativen Diskurs. Vdf ETH Zürich, part I

Renn O, Schrimpf M, Büttner T, Carius R, Köberle S, Oppermann B, Schneider E, Zöller K (1999) Bürger planen ein regionales Abfallkonzept. Akademie für Technikfolgenabschätzung in Baden-Württemberg, Stuttgart

Schmid G, Decker M, Ernst H, Fuchs H, Grünwald W, Grunwald A, Hofmann H, Mayor M, Rathgeber W, Simon U, Wyrwa D (2003) Small Dimensions and Material Properties. A Definition of Nanotechnology. Graue Reihe Nr. 35, Europäische Akademie GmbH, Bad Neuenahr-Ahrweiler

United States Senate (1972) Technology Assessment Act of 1972. Report of the Committee on Rules and Administration, 13 September 1972, Washington D.C.

Van Est R et al. (2002) The Netherlands: Seeking to involve wider publics in Technology Assessment. In: Joss S, Bellucci S (eds) Participatory Technology Assessment: European perspectives. Centre for the Study of Democracy, London

VDI, Verein Deutscher Ingenieure (ed) (1991) Richtlinie 3780 Technikbewertung, Begriffe und Grundlagen. Düsseldorf

Vorwerk V, Kämper E (1997) Externe Prozessbegleitstudie der dritten Phase des Bürgerbeteiligungsverfahrens in der Region Nordschwarzwald. Arbeitsbericht Nr. 97 der Akademie für Technikfolgenabschätzung in Baden-Württemberg, Stuttgart

Williamson RA (1994) U.S. Space Policy at the Office of Technology Assessment. In: Grunwald A, Sax H (eds) Technikfolgenbeurteilung der bemannten Raumfahrt, p 211–236

Part I – TAMI Main Results

1 The Practice of TA; Science, Interaction, and Communication

Danielle Bütschi, Rainer Carius, Michael Decker, Søren Gram,
Armin Grunwald, Petr Machleidt, Stef Steyaert, Rinie van Est

1.1
Introduction

As TAMI reflects on the activities of Technology Assessment (TA) institutions and their effectiveness, the central question seems to be: which methods should TA use in order to optimise impact? Although this question sounds quite easy, this paper shows that reflecting on the impact of TA methods is a very complex endeavour. The goal of optimising impact of TA activities requires a comprehensive reflection on TA processes, TA quality criteria and, the institutionalisation and mission of TA. In this paper we strive to provide a common ground for such a broad reflection.

Section 1 provides a general definition of Technology Assessment. Based on this definition, we develop a common framework in which to reflect on the relationship between method and impact in section 2. Elements in this framework that we consider are for example the institutional setting and the various phases of a TA project: situation appreciation, goal setting, project design, and project implementation. The following sections deal with those various elements. Section 3 distinguishes between various institutional settings in which TA projects are being designed and implemented. We discuss the institutional setting first because it influences the choices that are made in every step of the TA process. In sections 4 to 7 we describe the four phases that were introduced above. Section 4 discusses various dimensions or aspects that need to be considered when appreciating a situation. Attention is given to the issue dimension, the political dimension, social dimension, innovation dimension and the availability of knowledge. Section 5 describes various possible categories of goals. Section 6 deals with the project design phase and we introduce a method toolbox consisting of three classes of TA methods – scientific, interactive, and communication methods. In line with these three classes of methods three types of quality criteria are treated: scientific, interactive and communication quality criteria. These criteria refer to different sets of requirements that TA has to cope with, that is, TA has to comply with scientific and democratic demands and needs to have an impact on the political and societal debate. Section 7 deals with the project implementation phase. Finally, in section 8 we give an overview of the findings and draw some conclusions with respect to the relationship between methods and impacts.

1.2
TA definition

Obviously, the way by which TA methods have to be or could be mapped depends on the preceding understanding of TA. Even the scope of the notion "TA method" depends on the underlying TA definition. Therefore, the basic TA definition is decisive concerning both of the aspects: which methods have to be taken into account, and in which way they could or should be classified.

TA is a generic term that covers non uniform, and sometimes even contradictory approaches and activities and therefore it is difficult to define (Grunwald 2002). It is not obvious nor self-evident what the "common" should be, if one views technology forecasts, risk communication, problems of legitimisation, innovation funding etc. One of the main conflicts in the discussions on TA is based on the fact that everybody wants to have his/her own question concerning future technologies in the centre of the definition of TA.

However, in so doing they often pressure other aspects to the brink or even out of the TA-definition. This, of course, meets with stiff opposition from other TA-actors. However it is not very helpful continue with the discussion on this level. A definition of TA could be based on many different categories and if there is no consensus about these categories then not only is the definition not consensual but moreover there is no chance for a rational dealing with this problem. Possible definitions are:

- Definition via the tasks and functions of TA. Focussing on the contribution of TA to the social problem solving.
- Definition via certain special aims, e.g. early warning against technically induced risks (in the beginning of TA) or the stressing of innovation funding.
- Definition via the methods used. The kind of methods (e.g. scientific or participatory) used in TA are taken as defining categories.
- Definition via the subject-matter. What is investigated concretely by TA and on which aspects of technology is it related to?
- Definition via the addressees. Which persons, groups or social subsystems should be addressed and advised by TA?

It becomes obvious very soon that definitions on the basis of these criteria on the one hand will overlap but on the other will lead to diverged assignments. If one wants to have a common definition at all, and a common project on TA is a good reason for such a definition, then one has to identify what should and what should not to be taken into consideration, and a not too narrow definition according to this end is necessary. However, one should not forget that this definition has to be scrutinized in the context of the TAMI-project. The main criterion can be formulated in a means-end situation: Is the definition adequate for the purposes of TAMI?

The following proposal is based on the first mentioned category focussing on TA's contribution to the social problem solving:

> Technology assessment (TA) is a scientific, interactive and communicative process which aims to contribute to the formation of public and political opinion on societal aspects of science and technology.

This definition contains two substantial distinctions. Firstly, TA is dealing with contributions to public and political opinion forming and not with the decision making itself. TA offers knowledge, orientation or "approaches" to overcome social problems (e.g. unintended consequences, loss of confidence, problems concerning legitimisation). TA is neither able nor legitimated to solve these problems. This is part of societal decision making processes that take place in established institutions such as national parliaments. There is a difference between advising and shaping: TA is not shaping of technology but providing the knowledge for and advising to shaping of technology.

Secondly, science, interaction and communication are of crucial importance and form the three pillars of TA. *Science* provides knowledge about consequences of technology, conditions of implementation and mechanisms of controlling the technology development ("scientific methods"). The social relation to technology is characterized by problems of legitimisation, conflicts and loss of confidence. This is where TA offers and organises *interaction* between the opponents, the stakeholders etc. in order to overcome these problems. Examples for these "interactive methods" are risk assessment, mediation and participation of citizens. *Communication* is relevant for the distinction between "public" and "political" opinion forming. Obviously, "successful" communication in these fields has to meet different requirements. Therefore, communication is directly connected to "stay in touch" with the social surrounding. "Communicative methods" like newsletters, interactive websites, science theatres etc. are used to realize that.

The attribute "societal" refers to aspects of technology which are relevant for society. This includes ethical, economic, environmental, social aspects of technology. The attribute "technology related" refers to the question of which notion of technology is relevant in TA. Probably all attempts to develop a comprehensive notion of technology have to combine the substantial (artefact/tool) and the procedural (technical procedures) aspects of "technology". Often TA refers only and directly to artefacts/hardware and deals with the "more detailed analysis of the place of 'things' in chains of action" (Handlungsketten) (Wagner-Döbler 1989, p 25; in the same sense Ropohl 1979/1999). Relevant questions concerning this aspect are (a) how do human beings get the artefacts, (b) how do human beings behave relative to a artefact, (c) how do human beings get rid of the artefacts. Technical procedures and software are also part of the technical "world". Surgical operation techniques, programming and techniques of knowledge management can also be in the focus of TA. This procedural aspect of technology should not be disregarded. Objects of TA are in this sense of technology:

- ways of acting, in which technology is developed or produced;
- ways of acting, in which technology is used;
- ways of acting, in which technology is removed from the context of use (recycling, disposal).

The attribute "science related" takes into account that already scientific findings can lead to TA-relevant questions. The difference between scientific application oriented research and the early stages of a new technology/procedure are fuzzy (this is also the reason why "science and technology" is often used as fixed, inseparable expression). Therefore, TA-projects focussing on the "early warning" aspect, for example, are science related. Scientific research can also be on the agenda of TA, when they reach a certain order of magnitude. Huge high energy particle colliders, fusion laboratories are examples of that.

1.3
From method to impact: a complex relationship

TAMI, as a network that reflects on the relationship between method and impact not only needs a common definition of Technology Assessment, but it should be able to base its discussions on a common framework. Having such a framework is a necessary step in order to work with common notions and to understand the relationships between method and impacts.

As a first general feature, one has to acknowledge that the relationship between method and impact is very complex. Of course, TA practitioners must choose the right methods to have some impact. But, how to choose the right methods? What are the criteria to be used? Does one only have to pick the right methodology? Unfortunately, it is not that simple, the relationship between method and impact is much more complex.

Figure 1.1 tries to systematize the relevant variables intervening in the relationship between method and impact. According to this scheme, the possible impacts of TA activities depend on many variables, related to the TA institution itself, to the situation (or context), to the goals or objectives of the TA-project, to the selected method(s) and to the project management.

First of all, TA activities are always located in a specific societal, political or scientific situation, which is to some extent objective and given. A law may be under preparation, a large public protest might be under way, the main political actors may be trapped in conflict without a solution in sight, and so on. This situation is the starting point for the TA-project but is, off course, not static, it evolves. This means that a situation appreciation must be regularly undertaken. It is necessary to keep track in order to react or to adapt the project to the evolving situation. But, secondly, this also means that just the act of doing a TA-project, might already have a par-

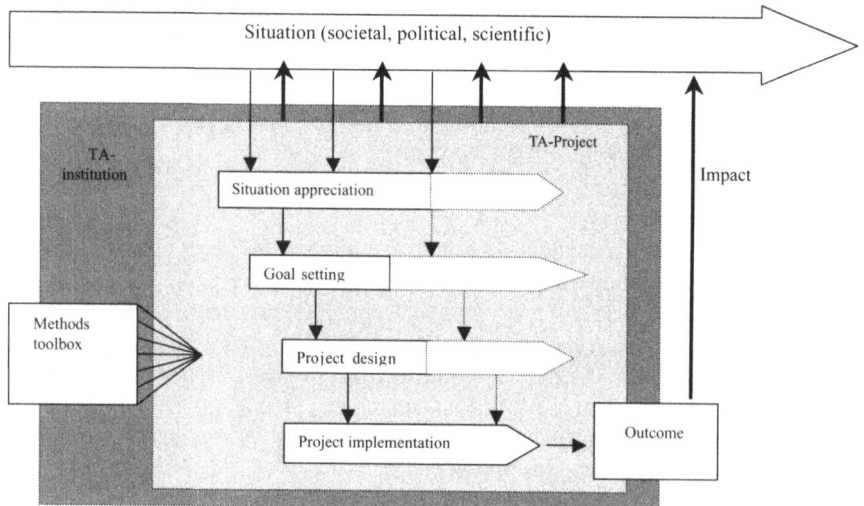

Fig. 1.1 From Method to Impact

ticular "impact" on the situation. For example, a major participatory event in the course of a project and the media attention that surrounds it might already influence decision-makers to take action before the report with the final conclusion is made public. Bringing together some experts around a topic might create new insights and new networks between these experts. If they take these back home, one can expect some impact on current situations. Therefore situation appreciation is an ongoing process during the TA-project and might lead to an adaptation of the TA-project design.

After the detailed situation appraisal is made, it can be used as the basis for goal setting. Before entering the design phase of the project, it is already necessary to think about the kind of (final) impact expected from the project. The type of impact one reaches for will influence the design (including the choice of methods) of the project. This "chain reaction" of situation appreciation, goal setting and project-design takes place after the initial situation appreciation at the beginning of the TA-project and may take place as a result of ongoing situation appreciation during the TA-project.

An institution's own TA "philosophy" will be of major importance in setting the goals. Each TA institution understands its mission in relation with its own specific organisational, cultural and political context. Some TA institutions situate them-selves more in the knowledge production area (giving input to policy making), while for others their mission lies more in the attitudes/opinion formation, where the process of participation and interaction is more important.

In order to operationalize TA activities, a set of methods is available (in a kind of "method toolbox") and can be picked up. Methods are manifold and can be used to collect data, provide knowledge, organise TA relevant communication, gain ideas for conflict resolution, and uncover the normative structure of technology conflicts, etc. Often, specific goals can only be reached by using a combination of different methods or by creating new ones. Therefore a TA-institution develops a project-design out of the methods with the highest potential of success. Obviously a proj-ect-design need not contain several methods, if it appears that one method, e.g. an eco-balancing is promising. The highest potential to reach the goals identified is only one criterion for the combination of the TA-methods. The project design also takes into account general quality criteria of TA like scientific reliability, fairness of interaction, etc. Moreover, the project design is influenced by the institutional set-ting: the mission of the institution, its tradition or history, its formal status, etc. This results in the fact that some methods of the method toolbox are preferentially used for project-design. For example, a TA institution whose mission is to gain interdis-ciplinary knowledge on new emerging technologies, might not have consensus con-ferences in its methods toolbox.

The realisation or implementation of the project design makes it a real world project. At this stage one no longer talks about ideal methods but about actions that are actually taking place. Here, variables such as project management, independ-ence, communication and other quality aspects are intervening. This phase is cer-tainly a crucial one: even though a TA institution makes a "correct" appreciation of the situation, sets the "right" goals, chooses the "right" set of methods and designs a "perfect" project, the whole project may fail if it has a bad project management or misses one of the above mentioned general quality criteria of TA.

Once the project is over, it produces outcomes, from which an impact is expected. As stated before, however, it is not only the final outcome which "produces" an impact, but also the activities during the process or the simple fact that the project is underway. According to this structure, the discussion of impacts related to methods needs to consider all phases of a TA project, from the idea to its realisation. This rather linear scheme can, however, be non-linear, as a constant and regular checking with the current situation has to be considered. In this respect, goals might evolve during the project, some additional phases might be added to the project design, and so on. However, the more the project is advanced, the more difficult it will be to modify it.

In the following we will use this structure to discuss the relationship between method and impact. Accordingly, all phases of a TA project – from the idea to its realisation – will be considered. We will discuss the different steps to better understand how methods lead to impacts. But before we describe the various phases, we will make some introductory remarks on the institutional setting in which TA projects are being shaped, since the specific institutional setting influences all the phases of a TA project.

1.4
Institutional setting

TA organisations have many different institutional settings, which vary from country to country and organisation to organisation. The specific institutional setting will affect the kind of activities they will set up and, consistently, the kind of impacts they might (and can) achieve. As such each institutional context leads to a certain extent to a "unique" project design. It can be said that the institutional setting of a TA organisation both enables and constrains the goal setting, the range of topics and methods chosen and type of impacts that are aimed at and/or can reasonably be expected. Moreover, the way quality control mechanisms are set up may differ from one TA institution to another. In this section we will make some remarks on the type of institutional settings, the target groups, and expertise with regards to issues and methods of TA organisations in various organisational contexts.

1.4.1
Types of institutional settings

Without trying to be complete the following organisational settings can be observed: scientific organisation, parliamentary TA bodies, consultancy agencies, and dialogue platforms. These various settings do not exclude each other. For example, a science-oriented institution may simultaneously fulfil counselling tasks.

In the case of TA as a scientific institution, the organisation is part of the academic system, where emphasis is put on scientific activities and scientific contribution to TA. Within parliamentary TA, the TA organisation belongs (or is subordinated) to a parliament. In such a setting, TA activities are mandated by parliament (individual members or commissions) and their results are intended to feed the legislative process. In independent public institutions TA is organised with the aim of counselling on science and technology issues. In this setting, the TA organisation has a certain freedom to choose its activities and projects, but it must take into account its potential clients in order to have any impact. In this setting, an appropriate appreciation of the situation is especially important.

1.4.2
Addressees or target groups

A crucial dimension to differentiate between institutions is the customer of the TA organisation. For example, parliamentary TA organisations are usually confronted with concrete questions from the parliament. Obviously the situation appreciation focuses around these questions, since the goals to be reached are more or less defined by parliament. Nevertheless a situation appreciation has to be carried out, because the selection of the method refers to both the goals and the situation appreciation. As a result of the latter additional goals might need to be taken into account. The chance for concrete (political) impact seems higher, since the parliamentary TA organisation reacts to demands of the parliamentarians and the addressee thus is willing to consider the end results and (in some cases) may even need these results to make decisions. On the other hand, being closely connected to parliament may also limit the play (i.e. scope of issues and methods) of a TA institute. TA organisations not directly connected to parliaments have a relatively harder job in selecting the topics parliamentarians are interested in and ensuring that the relevant target groups – policy makers, lay people, stakeholders – are interested in a specific issue.

1.4.3
Types of expertise on issues and methods

Another aspect concerns the variety of expertise on TA issues and methods that is available within a TA organisation. For example, one institute might focus on interdisciplinary expert discussions, while others specialise in participatory methods. These types of specialisation and former experiences of a TA institute influence the goal setting, enabling and constraining the various topics and methods used.

1.5
Situation appreciation

Technology Assessment addresses many different issues linked to technological advances and is thus confronted by a variety of questions to work on, with their own characteristics. TA might address questions related to nuclear energy, stem cells, privacy in the information age, waste management or sustainable development, just to cite some examples. Moreover, the same issue can lead to quite different questioning in different contexts. The way to tackle the issue might depend, for example, on the technological development, on the political context or on the general mandate of the TA institution. In this respect, a sound appreciation of the situation is an important step towards achieving any impact. The right appreciation of an issue and its context will help a TA institution to fix realistic and appropriate goals for a given project, and then use the appropriate methods in order to set up a TA approach able to realise these goals. It might be useless to launch a consensus conference on new technological developments without having, for example, previously considered what the technology can achieve (e.g. can xenotransplantation diminish the lack of available organs?) and their implication on law (is it authorized, do we need a revision of the existing laws, etc.?). On the contrary, a scientific study on a much discussed technology might be of less value than a participatory project. It is thus important to have the tools necessary to better apprehend the kind of situation a TA project is addressing, in order to implement the right method with the right goals and thus have a greater impact.

Practically, situation appreciation is part of the pre-phase of a TA project. Usually a kind of discourse analysis is used in order to draw a "map" of the ongoing debate. Who are the relevant actors and what arguments do they use? Which positions do they take against each other? Moreover, a media analysis, to investigate positions in the debate and the public resonance to a specific topic, can support such a discourse analysis as well as expert interviews and expert or stakeholder surveys, which should help clarify arguments and positions. The analysis of relevant documents complements the situation appreciation. Situation appreciation is of crucial importance, as is illustrated by the examples below,, because improper assessments will lead to projects contributing to solutions to irrelevant problems. However, due to practical reasons a balance between "quick" and "detailed" has to be realised. Moreover a monitoring of the situation is relevant during the whole project in order to keep track with changes to the environment.

In the following sections, an attempt is made to systematise the dimensions that need to be considered when appreciating a situation. We consider the (1) issue dimension, (2) political dimension, (3) social dimension, (4) innovation dimension, and (5) the availability of knowledge. This systemisation should be understood as a checklist and a rough guide of what should be taken into account. In order to point out the relevance of the situation appreciation for the following steps of "goal setting" and "project design" many references concerning and anticipating these steps are already made in the following sections.

1.5.1
Issue dimension

The first – and trivial – dimension is the issue a TA project addresses. Looking at the working programmes of TA institutions, we can see that the issues addressed are variegated: the themes are manifold and framed in different ways. In this context, when appreciating the situation, it might be important to be aware of the way the issue is framed. The following typology tries to systematise the different kinds of issues TA projects are addressing:

Technology oriented issue

This is the traditional – and maybe the more widespread – way of doing TA. The consequences of a technology already existing or in development or of a technological trend are analysed in an interdisciplinary manner or put under discussion among experts and/or laypeople. Many technology-oriented projects address questions related to new developments in biotechnology/biomedicine and in information and communication technologies. Other important issues are new materials, transportation systems, waste-treating technologies and energy supply techniques.

Examples: The Situation Appreciation in Practice

Drawing a Map of the Existing Debate on GMOs

In May 2002, the viWTA started with a project to give new impulses to the existing debate on GMOs in agriculture and food. Before making the methodological choices, a situation appreciation study was conducted. The situation appreciation study had four main goals:

- to list all relevant actors that are connected with the debate on GMOs in Flanders (social map). The authors of the study came up with more then one hundred organisations;
- to give an overview of the positions of the different actors. This work was based on a discourse analysis of websites, press releases, magazines, interviews,…
- to analyse the ongoing debate on GMO's in Flanders. Which arguments do the different stakeholders use, what do they think of each other. This part of the study showed very clearly that several stakeholders had a completely wrong idea of the positions of other stakeholders and especially of the position and opinion of the public.
- to map all existing and coming legislation on regional, national and European level that has to do with GMOs.

The conclusions of the study lead to the decision to organise a consensus conference and a stakeholders' forum in the course of the project. The consensus conference would aim for a clear and in-depth view of the opinion of the public. The stakeholders' forum (to be held at the end of the project) was meant to create an opportunity for stakeholders to exchange and discuss views and positions. The study also made it possible to inform and involve all the relevant stakeholders. In this way, it was possible to build up broad societal support for the project (Goorden et al. 2003).

Situation Appreciation of the Polarised Discussion on Cloning

In the autumn of 1997, the Dutch Minister of Public Health, Well-being and Sports asked the Rathenau Institute to contribute to the societal debate on cloning. The reasons were the public reactions to the birth of the cloned sheep, Dolly. Since political debate was high and the public discussion verypolarised the Rathenau Institute decided that an open and transparent process was needed to appreciate the situation and accordingly determine the project activities. To do this, the Rathenau Institute organised a hearing in the old conference room of the Lower House on March 26, 1998. At the hearing a panel of (former) Members of Parliament questioned researchers, representatives of biotechnology companies and interest groups, and ethicists about the state of the art of the technology, the possibilities of application, the arguments for and against certain applications of cloning and the reasoning behind these arguments. The hearing clarified three relevant issues for debate: cloning of stem cells, cloning of animals for the production of medicine, and animal cloning in animal husbandry. On these topics expert meetings, open to the public, were organised. Also two meetings were held to obtain more insight into the way various religious and political traditions deal with the ethical problems surrounding cloning. To further the public's input to the cloning debate a lay panel was set up. The panel could take part in all of the other activities organised by the Rathenau Institute and was also given the time and money to develop its own activities (e.g. questioning experts, visiting firms, etc.). (Biesboer et al. 1999)

Domain oriented issue

Technology intervenes in many domains of our life, like health, work, entertainment, mobility, etc. In this respect, the Technology Assessment community is also interested in evaluating or discussing how a certain domain of human activity is affected by new technologies. Classical examples of such projects are e-commerce or e-health. But more focussed projects can also be taken into consideration, such as road pricing or call centres.

Consequence oriented issue

The topic of interest might rely on certain consequences of technologies. In this context, the project will not mainly address the technology, but will put the emphasis on societal trends or changes that are technology related. Typical examples are projects addressing the questions of privacy, sustainable development, gender division, North/South relationship, etc.

In all these different types of issues the situation appreciation includes investigating the relationship between the specific TA issue and the specific context the TA project has to face. This usually will allow identification of the TA relevant aspects of issue and context in more detail. This fact, however, makes obvious that an individual TA-project touches all the above mentioned dimensions. A project that is domain oriented must, of course, consider the technologies involved and look at the consequences. A technology-oriented project must consider the domains that are affected by the technology under scrutiny and the consequences it may have. Finally a consequence oriented project must be aware of the different domains concerned with the problem and the technologies involved. Nevertheless this typology is helpful in order to identify where the project starts from and what its initial perspective is. This influences the goal setting of the TA-project and the project design.

1.5.2
Political dimension

Issues addressed by Technology Assessment are generally politically relevant. This relevance, however, might change depending on the stage of the policy-making process we are in, and the kind of political debate which is going on. In this section we distinguish between three phases: agenda setting, policy making and implementation phase. Finally, we discuss the situation in which the political debate and policy-making process is blocked.

Agenda setting phase

In this phase government has not yet officially addressed the issue. Still, an issue may be virulent among expert, citizens or interest groups. They may be aware of or sensitive to certain potential risks related to a new technology – for example environmental or health risks of new materials ("whistle-blower-effect"). On the contrary, some experts could have visions of unexpected utilities or chances of innovative technology. Anyway, in such cases there will be a high degree of ignorance and low awareness in general.

In this type of situation the goal of a TA project might be to assemble the knowledge available and to identify the areas of ignorance in order to be better capabilities of assessing the risk under consideration. In the case of proving the suspicion true, the "early warning" function of TA applies. The aim is to put the issue on the political agenda or to raise awareness in the political system concerning the issue under consideration. In other cases TA might lead to an early recognition of chances of new technologies.

Policy making phase
In this phase an issue is already on the political agenda, and at the stage where fundamental decisions have to be taken. In this case, a TA project might help to structure the debate, to make plural and comprehensive information available, to highlight and to assess the alternatives and the actions to be taken, to make the decision-making criteria transparent, in order to fulfil its task of supporting decision-making.

Policy implementation phase
In this stage of the policy cycle, there is a clear policy on the issue at stake (e.g. fostering e-learning), but the policy has still to be implemented. In this respect, TA might offer inputs on the ways to implement given policies related to technology in order to contribute to an efficient way of "domestication" or "embedding" the respective technology into society.

Political deadlock
Sometimes a debate on a certain issue is in a political deadlock; no solution is in sight. For example: the European debate on genetically modified food or the debates on nuclear waste disposal sites in many countries. Here, the role of TA in contributing to the management of conflict might be to overcome the blockade by communicative measures and by systematic analyses of possible alternatives. These can open new ways of thinking and new paths to overcome the deadlock-situation.

Interestingly, the phase and character of the policy debate might change in the course of the TA-project, especially for long and complex projects. As a matter of fact, when a project lasts several months – or even a couple of years – the chance that the political situation has evolved has to be taken into consideration and be integrated in the project design. In specific cases there might be a need to integrate special monitoring tasks into the project in order to prevent overlooking such changes or noticing them too late. This calls for an ongoing situation appraisal process. On the basis of this, the initial project design may be modified. Because political changes can happen suddenly, the project design needs some flexibility in order to react to such unexpected events (see section 6.3.3).

1.5.3
Social dimension

Another characteristic of a TA issue is the social dimension of the issue under consideration. We start the discussion of the social dimension with some general remarks about the role of values in TA. Next and related to that we will treat the following (to a certain extent overlapping) issues: public awareness, possible technology conflicts, and roles of various relevant actors and their relationships.

Value dimension

The value dimension is inherent to every technology as has been shown by TA in the last decades. However, there are differences in how deep ranging the relevance of values in a concrete case and situation is. The probability of the emergence of conflicts depends (among other factors) on the relevance of values. For example, technologies affecting deeply anchored values concerning the beginning and the end of human life, are often highly controversial in a pluralistic society. Other technologies, which might be considered under more economic issues (for example the substitution of classical materials by new ones), perhaps complemented by ecological aspects, can be judged in a more neutral way though they also show normative aspects. One of the tasks of a situation appreciation is, therefore, also to identify the relevant (perhaps hidden) values involved and their capability of leading to technology conflicts. Therefore a possible goal of a TA-project is the investigation of the TA issue with respect to the specific "framing" which has been chosen, and the search for alternative framing. The aim should be to make the particular values, that underlie the specific framing of an issue transparent, in order to make an open and transparent debate possible and to avoid bias in the TA.

Relation to the public

An adequate situation appreciation process should address the way the public is aware of or perceives a certain issue. Questions to be considered here are:

- Does the issue already raise interest within the public? Are there reports or discussions in the mass media (newspapers, TV, internet)? What role do the mass media play?
- How is the interest expressed if it exists? Is there fascination, rejection, mistrust against experts or against the political system, fear against the future, need for open debate, etc.? Are the chances or the risks of science and technology in the foreground?
- Who is leading the social discussion? Are large organisations (parties, churches, social movements) aware of the issue? Is there sufficient willingness among stakeholders and people affected to participate in a public debate?

Possible technology conflicts

With respect to the social dimension of a TA issue it is important to identify potential conflicts surrounding technology at an early stage. A situation appreciation in this respect should include the aspects of societal acceptance, power and communication. Relevant questions, therefore, are:

- Is there evidence for social acceptance problems of certain technologies? Could the debate even run into a blockade situation? How serious are positions of rejections to be taken?
- Has the debate so far been compatible to the requirements of social fairness or is there evidence for a severely unbalanced distribution of power in the public communication? Has the legitimisation of certain positions or even of democratic institutions in dealing with the respective technology been questioned?

Social roles and relationships

The design of a TA study may decisively depend of the assessment of the roles of experts, decision-makers and laypersons and their mutual relationships in the respective field. Has there been evidence for mistrust against experts and decision-makers? Does the expert dilemma (a situation in which experts are confronted with counter experts, science becomes politicised and the values behind science become explicit) apply? If these questions are answered with "yes" this will have an impact on how to conceptualise the TA project, for example concerning the use of participatory instruments.

In all these fields, the way an issue is discussed in the public might evolve during time. A scandal or a new scientific finding might deeply affect the social debate and thus the TA project on the issue. This shows the importance of keeping in track with developments in society and to be open for adaptation or reaction.

1.5.4
Innovation dimension

Similar to its impacts on the policy cycle, TA may play different roles at different stages of the innovation cycle. Some TA projects are rather prospective, in the sense that they explore technologies in the development stage, possible social practices that these technologies would imply and societal goals that are not yet discussed (take, for example, the TAB study on nuclear fusion research).

In other cases, TA projects will address questions related to existing technologies. Here, TA may focus on the shaping of technology. For example, do we want the internet to be better controlled or let as a free space? A TA may also consider the practical implications of the widespread use of a new technology, how to actualise this technology to the present situation. For example, what about the multiplication of in-vitro fertilisation in industrial countries?

Along the development path of a specific technology, TA has different entry points. TA relevant questions are different in the various phases of the innovation cycle, as well as there being different stakeholders and social groups involved. Accordingly, fitting TA questions to the development phase of the respective technology and to the corresponding decision-making requirements is an essential element of situation appreciation. Following the widely used model of the innovation chain, and adding the notion of an embedded technological system, we could identify the following different functions of TA.

Research and development in early stages

TA at the early R&D phase takes the function of a "science assessment". In the foreground are chances and risks of the developments, topics of public promotion of research and possible regulation needs.

Industrial research

Research for new products or processes under the rules of competition in the marketplace is much closer to having a direct impact on society. In this phase impacts and consequences of technology and ways of dealing with them are the main subject of TA.

Marketplace

Sometimes, it is once products enter the marketplace that the public discussion really starts. A prominent recent example of this is the introduction of GMOs on to the European market in the mid-1990s, which led to fierce political debate and eventually a temporary moratorium of GMOs in Europe. TA has responded to this situation with public participatory events to clarify what 'the public' actually thinks about the issue and why.

Widespread diffusion

Some technologies rapidly enter the market and have a high diffusion rate. For example the rapid introduction of mobile phones. This led to public discussion about health effects of mobile phones and its related infrastructure. TA's task is to acknowledge, discuss and clarify these voices or warnings and give them a proper place within the political debate. In many cases, for example asbestos and PCBs, early and even 'loud and late' warnings were ignored by decision-makers (Harremoës et al. 2001).

Embedded technological systems

Sometimes the limits of success of technological systems already deeply embedded in society come into view. An example of this is livestock farming in the Netherlands. The current goal is to move towards a more sustainable and animal friendly system. In such a case, established institutes are often too closely related to existing interests and are not able to break up existing practices and look for new ways. In the event of a looming crisis, TA may help to look for new ideas in order to innovate the system.

1.5.5
Availability of knowledge

TA has to provide knowledge and perform knowledge management. Knowledge generation in TA has specific difficulties because anticipatory and therefore hypothetical knowledge is required. The design of a TA study depends on the amount and quality of knowledge already available in the respective situation and on identified knowledge deficits and gaps. Therefore, an exploration of the availability of knowledge belongs to each TA pre-phase. Different points of departure in this field are:

- high-quality knowledge available, high degree of consensus among experts and scientists;
- high-quality knowledge available only in some relevant fields concerning the issue under consideration, with other areas of ignorance or high uncertainty;
- there is knowledge available about gaps in knowledge ("gewusstes Nichtwissen" – "acknowledged Ignorance"), for example due to some suspicions about technology risks without uncontested empirical evidence (for example this is the case for Electro-Magnetic Fields).

Obviously, the portion of TA related to knowledge generation and knowledge management – compared to the portion of discourse and communication – will be different in different situations – which then will have a large impact on the design of the respective TA project.

1.6
Defining the goals of a TA-project

After the appreciati of the general situation, the goals of a TA-project have to be defined. It is possible that a TA institution might decide to stop the process here, if it considers that the situation does not correspond to the kind it can contribute to (compare the paragraph on "institutional setting"). If it decides to continue, then many different kinds of goals may be possible. We can, however, systematize the kinds of goals around the following categories (clusters)[1]:

Scientific assessment
This is typically seen as the condition sine qua non of technology assessment. Different technical options have to be identified and assessed in comparison. This needs the gaining of knowledge about these technical options. Moreover knowledge about the societal, political, ethical consequences of these alternatives must be developed and presented in a comprehensive way.

Social mapping
If the situation appreciation leads to the identification of a social, political, ethical, etc. conflict the analysis of this conflict and its transparent description becomes a goal of the TA-project.

Policy analysis
This goal becomes relevant if a topic is already on the political agenda. Within the preparation phase of a political decision, the exploration of the respective objectives is essential. After a political decision, assessing the consequences resulting from this decision might be a goal.

Agenda setting
If a new technological development has not been considered yet by the politicians or by the public one of the project goals might be to raise awareness for the technology and its consequences, in order to put it on the political agenda or to stimulate a public debate about it. It might be sensible to support this process by illustrating the relevance of the technology through the development of visions and scenarios about it.

Mediation
After realisation that a blockade situation exists, a TA-project might aim to overcome this situation. This can be realised by stimulating the actors to go through a

---

[1]   These goals of TA-project can be compared or are mainly identical with the "roles" of TA identified from the impact analysis perspective (cp. chapter 2). Therefore we choose the categories which can also be found in the typology of impacts, examples for the roles can be found in the annex I.

self-reflection process or by the development of bridge-building alternatives. The more general goal behind this is to enable the opponents to overcome the blockade.

Restructuring the political debate

TA-projects might aim to influence the ongoing political debate. One goal might be to increase the comprehensiveness of the debate, or to evaluate the existing policies through a new discussion process or to reach for a kind of democratic legitimisation of political decisions.

Initialize new R&D policy

Another goal of a TA-project might be to develop recommendations for new topics on the research agenda. For example if an identified problem causes the need for further research. Also the development and assessment of various options might lead to recommendations for a re-orientation of research policy.

New decision making processes

The situation appreciation might result in the finding that a particular situation needs new decision preparation procedures. It might be a goal then, to recommend an alternative way for governance. In a broader sense the initialisation and intensification of a public debate also might be a possible goal of a TA-project.

New policies

One possible goal of a TA-project might be to recommend concrete policy activities. A new technology might lead to the fact that existing laws need extension or modification. Another possible goal is the evaluation of different technology policy alternatives.

Of course, a TA project might address several of the above mentioned objectives. For example, a TA project might aim at improving knowledge on a new emerging technology and informing the public about it. Moreover, this list is not claimed as to be complete. Goals have to be defined in accordance to the situation the TA institution has to cope with. Imagine a situation which, according to the above mentioned dimensions,[2] is technology oriented, the issue is not yet officially addressed by government, there is low public interest so far and the technology is still in the labs of applied research on a prototype level. The facts available about this technology are mainly disciplinary, i.e. provided by the developers of the new technologies. Therefore one goal of a TA-project could be to analyse the social shaping of this technology, i.e. how the technology becomes a part of our culture. In another case the appreciating of the general situation might show that the issue is in a political deadlock, the public mistrusts the decision makers and that the questions being asked are related to existing technologies. The social discussion is more than lively, influenced by the media, and NGO's have already formulated their positions. In this situation, it might be relevant to bring the opponents together, involving the relevant NGO's and stakeholders, and to organise structured discussions in order to overcome the blockade. But it might also be of interest to start an interdisciplinary

---

[2] See the previous paragraph "Situation appreciation".

research project in order to acquire arguments to get the debate down to a more rational tier. Both can of course be done in the same project[3].

1.7
The project design

Based on the specified goals, a project design has to be defined. This project design is essential in order to achieve the goals, and consequently some impact. A TA-project design is developed out of the different methods considered to be most promising to fulfil the forthcoming tasks. In the first part of this chapter, we will give an overview of the "method toolbox". Through the years, the TA community constantly tested new methods, improved existing methods and widened its set of methods. The selection of methods will need to be in coherence with the defined goals of the TA-project. In the second part, we will highlight the most relevant principles in selecting methods.

But a project design is more then just choosing the appropriate methods. In this phase, it is important to think about the quality requirements each project has to meet. Since we defined TA as a scientific, interactive and communicative process, the necessary steps need to be taken to guarantee the quality of these three pillars of TA. In the third part, we will describe the different quality criteria that have to be taken into account.

1.7.1
The "method toolbox": three classes of TA methods

According to our structure "from method to impact" (fig. 1.1), we see a TA-method as a structure, composed out of multiple elements (techniques, instruments), that serves the scientific, interactive and communicative goals of a TA-project, picked up in a "method toolbox". This method toolbox contains all kind of methods which can be justified as to be helpful to reach a specific outcome and, thus, the set goals (or sub-goals) of the TA-project. This outcome may be a product as such, for instance a study, or a set of future scenario's and so on, or it might be the process (or consequence of it) itself, for instance the creation of a network, a debate in parliament, etc. Based on this definition, a TA project is seen as a combination of different methods, which have to be structured in a "project design".

When looking at TA projects, one sees that many methods are used and it might be difficult to describe them in detail. Nevertheless, some methods constitute the core of the TA method toolbox, whereas new methods are constantly being tested. In fact, the TA method toolbox expands with time and with new institutions joining the TA community.

---

[3] An example for this situation could be a TA-project dealing with the climate change problem. On the one hand it is necessary to scientifically scrutinise the variety of climate models, which are the basis for the prognosis of future climate change. A scientific questioning would also consider economical, legal, ethical consequences of the different options. On the other hand these options will influence the every day life of the citizens, for example when the reduction of CO_2 is aimed at, which needs reorganisation of every day habits. This justifies participatory TA in order to find the "factual accepted" solutions.

Examples of TA Methods

The Delphi Method

Delphi involves a survey of experts. Delphis focus on forecasting technological or social developments, helping to identify and prioritise policy goals, or determining expert opinion about some aspect of affairs that cannot be measured directly by conventional statistical means. Delphi was designed to provide the benefits of a pooling and exchange of opinions, so that respondents can learn from each others' views, without the sort of undue influence likely in conventional face-to-face settings (which are typically dominated by the people who talk the loudest or have most prestige). Each participant completes a questionnaire and then is given feedback on the whole set of responses. With this information in hand, (s)he then fills in the questionnaire again, this time providing explanations for any views they hold that were significantly divergent from the others'. The explanations serve as useful intelligence for others. The idea is that dissenting views that are based on privileged or rare information can thus be weighed up by the entire group.

Modelling and Simulation

Within the project "Global sustainable development – perspectives for Germany" (Coenen/Grunwald 2003) input-output analyses on the basis of statistical data available have been used for a macroeconomic environmental-economic simulation, by means of which various scenarios of future development have been computed (using and furthering the model "Pantha Rei" developed at the University of Osnabrück). In this manner, sustainability-relevant developments including the implementation of new energy technologies have been projected into the future in order to get anticipatory knowledge about the future development of sustainability deficits and problems, according to different overarching scenarios. The future development of sustainability indicators has been calculated for the societal fields of activity (like mobility and traffic, housing and construction, nutrition and agriculture) as well as macroeconomic interrelations have been taken into account system-analytically. Furthermore, the consequences of possible political strategies and measures towards sustainable development have been estimated. In this way, modelling and simulation can help recognise and assess future developments – of course under the condition of uncertainty.

Consensus Conferences

A consensus conference is a public enquiry centred around a group of 15–30 citizens who are charged with the assessment of a socially controversial topic. These lay people put their questions and concerns to a panel of experts, assess the experts' answers, and then negotiate among themselves. The result is a consensus statement that is made public in the form of a written report directed at parliamentarians, policy makers and the general public that expresses their expectations, concerns and recommendations at the end of the conference. The goal is to broaden the debate on a given issue, include the viewpoints of non-experts, and arrive at a consensus opinion upon which policy decisions can be based. Consensus conferences usually have two closed, preparatory weekends (only the lay panel and their coaches) and a 3-day intensive programme that is open to the public.

Scenario Workshops

Scenarios consist of visions of future states and paths of development, organised in a systematic way. They can be either extrapolative or normative, but should enable participants to build internally consistent pictures of future possibilities and are useful for envisaging the implications of uncertain developments and examining the scope for action. Scenario analysis engages a group in a process of identifying key issues and ▶

then creating and exploring scenarios in order to explore the range of available choices involved in preparing for the future, test how well such choices would succeed in various possible futures, and prepare a rough timetable for future events. The method was designed to challenge the mind-set of participants by developing scenarios of alternative futures in order to understand how the world might unfold and how that understanding can be used in strategic planning. Tools & techniques often used: brainstorming, group support facilities such as the Group Decision Room software, models.

Science Theatre

It is not easy to address complex scientific issues in a for (young) people understandable way. Science theatre can bring up dilemmas and question around scientific topics in a personal or emotional and comprehensible manner. The play emphasises societal issues surrounding a scientific topic. After the play, the various actors stay in their role and enter into a discussion with the audience, which is led by a professional discussion leader. Of course several preparations are made before the science theatre combined with ethical debate can be performed. A scenario writer first has to write a script. A reader's commission of scientists advises the author about the scientific topic that is central to the script. After rehearsals the theatre company that performs the play can start its tour. When young people are the target group, normally some hundred secondary schools are visited. To help teachers in preparing their students for the play and the ensuing debate, a teacher's manual is written that introduces the play, the various involved issues, positions and arguments.

Looking back to the TA history, we can trace back the first expansion of the method toolbox when classical or scientific methods have been supplemented with participatory or dialogue methods– which broadly corresponds to the exportation of TA from the USA to Europe. Scientific and dialogue methods complement each other well and they are now considered as current TA practice. This extension of methods has also led to a broadening of the kind of quality criteria a TA project should meet. Besides scientific quality criteria for the output of the TA project, quality criteria for the TA process have been formulated (part 3 of this chapter).

Over the last few years, however, the TA community has become to realise the importance of communicating to the outside world about the TA approach and methods, process and output. Although communication tools have been always used in TA projects, these are now considered by some TA institutions as a true integral part of doing TA, thus belonging to the TA project design. In other words, so called *communication methods* are under development now. These "new" kinds of methods complement scientific and interactive methods in current TA practice[4], so that a TA project can be seen as a combination or mix of instruments from three classes of methods, i.e. scientific, interactive, and communication methods.

From this general history of TA, we can deduct three classes of methods characterising the "method toolbox":

Scientific methods
Scientific methods are developed in disciplines of natural or social sciences applied to TA problems, in order to collect data, to allow prediction, to make quantitative

---

[4] As well as there is a grey zone between scientific and participatory methods, there is also overlap between participatory and communication methods. Nevertheless, to conceptualise the meaning of communication it is relevant to distinguish between these three classes of methods.

risk assessments, to allow for the identification of economic consequences, to investigate social values or acceptance problems, to enable for eco-balancing. This class of methods includes:

- Delphi method, expert interviews (for collecting expert knowledge);
- Expert discussion;
- Modelling, simulation, systems analysis, risk analysis, material flow analysis (for understanding the socio-technical system to be investigated);
- Trend extrapolation, simulation, scenario technique (for creating knowledge to think about the future);
- Discourse analysis, value research, ethics, value tree analysis (for evaluating and uncovering the argumentative landscape);
- Etc.

Interactive methods

Interactive, participatory or dialogue methods are developed to organise social interaction in order to make conflict management easier, to allow for conflict resolution, to bring together scientific expertise and citizens, to involve stakeholders in decision-making processes, to mobilise citizens for shaping society's future, etc. This class of methods includes:

- Consensus conference;
- Expert hearing;
- Focus group;
- Citizens jury;
- Future search conference;
- Scenario workshop;
- Perspective workshop;
- Etc.

Communication methods

Communication should be seen as a two way process. On the one hand communication methods are used to communicate the corporate image of a TA-institute, the TA approach, the TA process and product to the outside world in order to increase the impact of TA. On the other hand communication is an important feature for the TA-institute to keep in touch with the outside world and to keep track with reality. This class of methods includes:

- Newsletter and focus magazine;
- Opinion article;
- Science theatre;
- Video presentation;
- (Interactive) websites (e.g. local questionnaire, debate forum, video, ...);
- Networking;
- Accompanying;
- Dialogue conferences.

1.7.2
Selection of the TA-methods

After the situation appreciation and the goal setting the development of a TA-design starts. As highlighted above, it is composed out of different methods. The main criterion for the method composition process is the development of a TA-design, which can be justified as the one with the highest potential for reaching the identified goals. However, in addition to this main criterion, more general aspects concerning "best practices" in TA also will be taken into consideration. Therefore the selection of the relevant methods to develop a concrete TA-approach is based on both the justification for reaching the goal and the necessary condition to reach high quality TA-results. Here also, the need to keep pace with the evolving societal, political and scientific situation has to be taken into consideration.

Justifying the selection of the relevant methods is done by referring to the goal setting after the situation appreciation. In the case of a project aimed at creating knowledge, one could start with identifying the relevant scientific disciplines, selecting the experts out of these disciplines, organising and moderating the expert communication process, etc. Another option would be to establish where knowledge is needed and launch an interdisciplinary study aimed at collecting knowledge via bibliography research, interviews, etc. If the situation appreciation shows that the different involved actors are in conflict or a deadlock, one would start with identifying possible stakeholders, perhaps identifying citizens concerned by the new technology under scrutiny, organising and moderating the communication process, selecting ways how the scientific knowledge is transferred into the discussion process, etc.

Referring to the three classes of methods in the so-called "method tool box", it has to be underlined that there is no need to combine methods out of the three classes in order to design a "complete" project. "Complete" in the sense of most promising refers to the situation appreciation and the following goal setting, only. There may be cases in which pure scientific inquiry or an individual newspaper article seems to be most effective. But it is also not the case that a project design should contain only one method out of each class. An expert hearing followed by a citizens jury or a Delphi process combined with an expert discussion might be identified as the most promising project design. However, as mentioned above, the project design is developed referring to more general quality criteria as well, for example to transparency, inter- and transsubjective validity, neutrality, etc. In the following examples these general quality criteria are described with reference to the three classes of TA-methods.

1.7.3
Quality criteria

Designing a TA project is not only about choosing the appropriate method or about planning steps, but it is also about taking into account quality criteria related to any TA activity. Such quality criteria are of crucial importance in order to achieve the set goals, i.e. to have the intended impact. Due to the specific nature of Technology Assessment, different types of quality requirements referring to the project design

Examples for Combinations of TA Methods

Project "Towards a societal agenda for food genomics"

The specific application of genomics to research in agriculture, foodstuffs and nutrition is called food genomics. Although many scientists expect that the societal impact of food genomics research will be huge, the understanding and awareness of (possible) social and moral aspects is still in its infancy. In the Netherlands, it was mainly policy makers and scientists from public and private research institutions who defined the direction of genomics research; other societal actors were sparsely involved in the debate. To change this and to contribute to the understanding and awareness of societal issues related to food genomics, the Rathenau Institute set up in 2002 the TA project "Towards a societal agenda for food genomics", which comprised of several types of activities. A scientific study was done to map food genomics research in the Netherlands, and its potential applications. Five social essays were written to explore the societal impact of food genomics on the entire food chain. The drafts of these essays were discussed in an expert workshop that confronted the essayists with scientists involved in food genomics research. The final versions were discussed during a working conference with a broad audience, consisting of policy makers, business people, scientists and members of societal organisations. The results of the studies, the essays and the discussions were presented during a public hearing, in which five parliamentarians questioned ten scientific experts and representatives from industry and societal organisations (van Est et al. 2003).

Co-operative Discourse

The model "cooperative discourse" is a hybrid model of citizen participation named by Renn and Webler (1998) consisting of three consecutive steps: (1) The identification and selection of concerns and evaluative criteria is accomplished by asking all relevant stakeholder groups to reveal their values and criteria for judging different options. To elicit the values and criteria the technique of value-tree analysis has proven appropriate. The resulting output of a value-tree process is a hierarchically structured list of values representing the concerns of all affected parties. (2) The identification and measurement of impacts and consequences related to different policy options is the following step. The evaluative criteria derived from the value-tree are transformed into indicators by a research team. Experts from varying academic disciplines and with diverse perspectives on the topic of the discourse are asked to judge the performance of each option on each indicator. For this purpose, a modification of the Delphi method has been developed and applied. The desired outcome is a specification of the range of scientifically legitimate and defensible expert judgments and a distribution of these opinions among the expert community with verbal justifications for opinions that deviate from the median viewpoint. (3) The last step is the evaluation of potential solutions by selected citizens as jurors and representation of interest groups as witnesses. These panels are given the opportunity to evaluate and design policy options based on the knowledge of the likely consequences and their own values and preferences. The participants are informed about the options, the evaluative criteria, and the performance profiles. The representatives of interest groups and the experts take part in the process as witnesses; they provide their arguments and evidence to the panels who ultimately decide on the various options.

The model has been applied with several modifications to studies on energy policies and waste disposal issues in Germany, Switzerland and USA.

can be established. More precisely, as a scientific, interactive and communicative process, Technology Assessment has to cope with requirements referring to:

- the scientific quality of the project;
- the design of the interaction process about technologies and their consequences;
- the dissemination of the work and of the results to the outside world.

Scientific quality criteria

Technology Assessment (TA) is rooted in the scientific activity. For its contribution to be of any value, it has to meet certain scientific quality criteria. TA has to provide knowledge about scientific and technological developments, about their consequences and about ways to cope with their possible risks and maximise their advantages. In this respect, TA projects must be scientifically robust. Moreover, due to the very special nature of TA looking at technologies embedded in their social and political environment, TA projects must meet quality criteria to cope with a multi faceted setting. We will discuss this point first.

Interdisciplinarity

The necessity for interdisciplinary research results from the issues that TA considers (Decker 2001). Only in very rare cases is one individual discipline able to deliver scientific knowledge in a comprehensive manner. In all other cases, teamwork between the relevant scientific disciplines is necessary, in order to overcome the scientific boundaries and thus contribute to the solution of a concrete problem. One has to identify and to justify how the disciplines should organise their work in order to reach compatible results and how deep or broad a scientific analysis should be in order to meet the relevant aspects of the project. The organisation of an interdisciplinary scientific project is a complex enterprise, which needs:

Disciplinary statements on the topic

Notions and underlying disciplinary assumptions have to be identified in order to find a common language, in which apparently identical notions mean the same thing.

Determination of the project relevant framework

A (sometimes hypothetical) predefinition of what is relevant to answer the question/problem (the border-line/cutting-off of the project/problem) has to be made in order to reduce reality to a manageable volume.

Combination of the scientific perspectives

The cross-correlations that are relevant to solving the problem on the agenda have to be identified. In addition, how these interdisciplinary cross-correlations can be reached in detail needs to be considered. Which questions from other disciplines have to be answered? How deep and how broad should the respective input from each scientific discipline be? It is a kind of "pragmatic compatibility" (Decker and Grunwald 2001) one reaches for.

Scientific reliability

High quality of the scientific input is a necessary condition for TA. The relevant trends in each discipline must be represented and the statements must be on the actual state of the art. A TA-project which obviously neglects a relevant scientific perspective will be labelled as "biased" before the results have been presented. Moreover, as TA is usually addressing controversial issues, it often faces a situation where experts interpret scientific facts in different ways, so that their opinions differ. It can also face a situation where experts "mix up" their scientific statements with their personal (political, economical, etc.) preferences and strategies, but might disguise their personal preferences with scientific arguments. All these elements make scientific reliability a decisive and necessary quality criterion for any TA project.

In order to achieve scientific reliability, the following procedures can be implemented:

Extended peer reviews

Peer reviews are a well accepted tool to evaluate disciplinary research. Second and third expert opinions have to be taken into account, especially when contradicting thesis are followed within a scientific discipline. In the case of a TA project, the peer review has to be extended in order to match the interdisciplinary requirements. This means that the peer reviewer will need to evaluate if the relevant cross-disciplinary questions concerning the issue have been asked and if the answers meet the relevant aspects of the respective disciplines. This evaluation process has to be taken seriously to avoid a kind of "science light"[5].These extended peer reviews can occur at the end of the project (classical peer review) or can be organised in order to "accompany" a project (for example, in a kind of "accompanying group" where experts of the involved disciplines would be invited).

Expert confrontation

When a situation arises where scientists hold different values or interpret facts differently, a direct confrontation of the experts may resolve this so-called expert dilemma (Nennen and Garbe 1996). With such a confrontation, and given that "good will" is present, experts should be able to identify their different underlying assumptions and weighing of arguments.

Interactive quality criteria

Regardless of whether TA wants to contribute to a public debate or a decision making process, the chances of having an impact are greater if the project can prove a high degree of legitimacy. Looking at the different stages of development in a TA-project (situation appreciation, goal setting, project design, project implementation, etc.) we cannot negate that the legitimacy of each stage depends very much on how the existing

---

[5]   "Science light" refers to the fact, that in interdisciplinary discussions in which each discipline is represented by one expert without evaluation, everything that this experts mentions is taken without question.

Examples of Realizing Scientific Quality Criteria in Practice:

Evaluation Loops within Expert Discussions

The so-called project group principle (Decker and Neumann-Held 2003) of the European Academy GmbH is in general an interdisciplinary expert discussion accompanied by several interdisciplinary evaluations. Since these evaluations take place during the whole discussion process they can be described as "loops" which the discussion processes has to pass. At the beginning of the project an interdisciplinary expert group (the scientific council of the academy) scrutinizes the work programme, i.e. the main questions to be answered, the scientific disciplines identified to be relevant, the way it is planned to overcome the disciplinary boundaries. Since "corrections" of the main focus are hard to do during the discussion process a Kickoff-meeting is organised at the beginning of the project to put the perspective of external experts on record. During the discussion process a Midterm-meeting takes place during which external experts comment on the work in progress and get the chance to ask for additional aspects not taken into account by the expert discussion so far. Finally, another expert group (the scientific council) scrutinizes the whole discussion process taking into account all "interim-evaluations". This group makes the final decision of whether the project is acceptable or needs further revision. By these evaluation loops it might be possible to overcome the so-called expert dilemmas like expertise and counter-expertise or the covering of personal interests by scientific statements by these expert confrontations and extended peer reviews.

The Group Delphi

The CTA Baden-Württemberg has developed a method called the group Delphi (Webler et al. 1991; Renn et al. 1999). It is similar to the original Delphi exercise but based on group interactions instead of individual written responses. During a group Delphi all participants meet face to face and make the assessments in randomly assigned small groups of three and four. The groups whose average scores deviate most from the median of all other groups are requested to defend their position in a plenary session. Then the small groups are reshuffled and perform the same task again. This process can be iterated three or four times until no further significant changes are made. At the end of a Delphi process one gets either a normal distribution of assessments around a common median, a two- or three peak distribution (signalling a majority and one or more minority votes) or a flat curve (which means that knowledge is insufficient to make any reliable assessment). The advantage of Delphi is that a serious effort has been invested to find the common ground among the experts and to find the reasons and arguments that cause differences in assessments. The disadvantage is that Delphis depend on the quality and completeness of the expertise and information brought into the process. There are several positive experiences with Delphi processes, in particular with group Delphi.

differences in social values are integrated. In this respect, social fairness, process fairness, argumentative quality and transparency are key quality criteria for TA projects.

Social fairness

The selection of the institutions or persons who will contribute to the project, i.e. interact with each other, is of crucial importance. According to the situation appreciation, the set goals and the chosen method, a TA project might be based on an interaction between experts, between stakeholders or between scientists and ordinary citizens, just to cite some examples. It is thus very important that, with respect to the aims and design of the project, the appropriate groups of participants are invited. For example, in the case of a project aimed at gaining knowledge about a new technology and its consequences, it might be appropriate to invite experts only. The question will thus be which disciplines or scientific institutions need to be represented. In a project aimed at solving social conflicts, it is crucial to invite stakeholders or citizens. The task is to identify whose interests are at stake and which persons have the legitimacy to represent these interests. This can be realized either by democratic elections, by authorization of groups by concerned people or groups or by formal procedures of equal opportunities (ex. random selection).

Process fairness

In a TA project, procedural fairness has to be implemented, i.e. that fair rules have to be established in order to allow all participants to be heard and considered. More precisely, process fairness applies to:

- *Meet dialogue standards*

  All invited participants should be allowed to have an input and to be listened to. More precisely, independent of their charisma, knowledge on the issue or social status, participants have the same rights to make assumptions, express opinions and formulate proposals.

- *Establish democratic decision procedures*

  To obtain fairness during a TA project, decisions on the content should be of collective nature, which implies a voting modus. The decisions can be taken with simple majority votes, absolute majority votes, multiples majorities or consensus. Decision procedures can be the same for the whole dialogue or be established from case to case.

- *Set an agenda*

  It must be clear to all participants what points will be discussed. Participants should also have the opportunity to modify the agenda, or even to set the agenda themselves. But finally decisions on the agenda or the procedure must be consensual.

Support towards the competence of the participants

This means that the information and knowledge handed over to the participants is understandable, i.e. all participants should understand all used concepts and definitions. If necessary, a translation should be available. Not only is attention for the transfer and acquisition of knowledge needed, but also for the acquisition of interaction skills by the participants. This concerns talking skills, group-working skills, etc.

Transparency of the interactive processes

Often Technology Assessment is carried out in very controversial fields, where different kinds of values or interests are at stake. This means that power relationships are present: interest representatives or holders of certain kinds of values will try to get their point of view in to the discussion and in to the final recommendations. Arguments might be a way to achieve this goal, but many other means are available, like persuasion, negotiation, threats, etc. Technology Assessment has to cope with this reality: when entering highly controversial fields like genetic testing or e-surveillance, TA is entering a political field with all its power battles. TA practitioners must be aware of these power relationships, and especially make them transparent, so that the participants themselves know who is representing whom. But transparency is even more important for those who are not part of the interaction, but might be concerned by the results of the TA project. Only with transparency can they understand how the project arrived at a certain result and thus check whether the process has been biased or not. In this respect, transparency means:

- *Transparency about interests and values*

  It is desirable, that each participant declares their interests. For example each participant should declare their institutional affiliation, its function or their job. It is also important that other mandates (in politics or in associations) are explicitly declared, as far as they are connected to the discussed issue. Additionally it should be made transparent whether statements are based on scientific knowledge or rather on values or personal interpretations.

- *Documentation about all steps*

  Each step and decisions of the interaction should be documented, for example in minutes or reports. The public and interested persons should have access to these documents. In some cases, the most important steps and decisions could even be made available in a kind of annex to a final report or on the World Wide Web.

Argumentative quality

At each stage within a TA-project claims are put forward and arguments are used. For instance: every TA-organisation needs to motivate every time the choice for the topic or focus for a new TA-project; in a dialogue between experts and members of a lay panel, knowledge, uncertainity, opinions and beliefs are confronted among the participants. Probably the statements of the experts will be more 'scientific' in comparison with these of the lay panel, but they all need to be justified with arguments. These arguments have to meet certain standards, that is have a certain syntax. According to the argumentation theory, a claim must be based on ideas (grounds), must explain why these grounds lead the speaker to conclude the claim (warrant), must give reasons why the receiver should believe the warrant (backing), and must contain qualifiers defining the scope of application of the claim ("if", "as long as", etc.) (Toulmin 1958, Freeman 1991).

Of course, it is impossible (and even not desirable) to check the syntax of every individual argument used in the course of a complete TA-project. This would involve interruptions to every discussion between people, to check all arguments in every document made, to prove every statement made by each expert. But, it must be possible to validate the quality of the arguments at decisive or important moments in a TA-project: in the publication of an expert report, in the recommendations of the laypanel in a consensus conference, in outlining the policy options in an advice to a parliament and – why not – in choosing a method for a specific TA-project or in selecting collaborating experts. The argumentative quality is probably the most important aspect if one seeks to influence the policy making process. If a TA-project is short on argumentative quality why should policy makers listen to it?

Examples: Interactive Quality Criteria in Practice

Process fairness through a declaration of intention: the Swiss "PubliForum" example

For each of its consensus conferences (called "PubliForum"), TA-SWISS sets up a "declaration of intention", in which the aims and the procedure of the project are stated. This declaration of intention contains also a list of standards for a "fair and transparent PubliForum", with statements such as "every opinion and position has to be considered by participants"; "organisers should clearly indicate how the PubliForum results will be used", etc. Moreover, special rules for each involved actors (laypanel, experts, facilitator, organisers, accompanying group) are formalised. These rules specify the tasks, rights and duties of the respective groups or persons (Joss and Brownlea 1998).

Social fairness through participant selection: the Danish "Future search conference" example

In 1998 the Danish Board of Technology organised a "Future Search conference". The method is suitable for breaking the ice in a (political) locked situation and in this case the topic in Denmark was about traffic in Copenhagen. The Future Search guides the participants through a process where based on history and analysis of the present day situation they create common pictures of the future and make concrete action plans to reach these goals. Therefore, it is of vital importance for the success that the "whole system" is present through the three-days conference, meaning that all stakeholder groups (e.g. environmental groups) and groups with influence on decision-making (e.g. politicians) participate to ensure the foundation for action afterwards. In the Danish case eight groups were identified: Business/economic life, politicians, officials, experts, environmental organizations, organizations of cars and road traffic, citizens using cars, citizens using bicycles and public transportation (Weisbord and Janoff 1995; Gram 1998).

Ensuring argumentative quality by making explicit, discussing and checking various policy options and problem definitions: The Gideon Project on sustainable crop protection

In the mid-1990s an evaluation was planned of the Dutch governmental multi-year plan for drop protection. In order to provide the members of Parliament with information on opportunities and threats for achieving sustainable crop protection the Rathenau Institute started a project, dubbed Gideon. The project was set up as an 'interactive' TA (Grin et al. 1997) consisting of various activities that involved many stakeholders. Various stakeholders had different ideas on what the goal of reducing the dependence on pesticides meant and how this objective could be obtained. A key characteristic of the methodology, therefore, was to make explicit and discuss both policy options and problem definitions. During all activities (amongst others in-depths interviews, future-oriented workshop, work conference, open day) it tried to include a large variety of perspectives. Moreover, the support within the field for the findings and results as formulated by the project team were constantly checked through interviews and debating events (van Est et al. 2002).

Communication quality criteria

These criteria relate to communication of the TA approach, process, and product into the outside world, for example to relevant target groups. Below, we formulate some communication rules or impact stimulation criteria.

Flexibility related to the ongoing debate

Flexibility is necessary during the project implementation and dissemination phase as the original situation appreciation might change. Such changes might concern the political agenda (e.g. a law is postponed or advanced), the public debate (e.g. a "scandal" raises interest for an issue, an issue has a "media peak" for a few weeks) or the scientific community (e.g. new findings). In order to achieve an impact, the TA project has to adapt to these changes. To achieve flexibility one should consider the following issues.

- *Adaptability*

  The project design must allow space for modifications during the performance. Highly complex and interdependent project phases might endanger the ability for adaptation. Such projects are difficult to adapt to new situations, as a change in an initial phase might have consequences on the subsequent phases.

- *Flexible institutional procedures*

  Project and institutional procedures (e.g. quality control procedures) must be treated with openness. For example, in many TA institutions a formal decision of a Board must be taken before publication. Such procedures might be slow and, if necessary, ways have to be found in order to be in tune with a new political, societal or scientific agenda. In justified cases rules are there to be broken.

TA institutes and practitioners should be open minded and focus on achieving impact

TA institutions and project staff must be open to short-term changes. Professionals must be open-minded and engaged in impact achievement.

Opportunities to create impact after the project should be looked for

Changes in the situation might happen once a project is over. In this case, it might be useful to actualize a project. This does not only mean to "sell" a finished project report, but maybe to amend the project to the new situation.

Keeping track with social, political and scientific reality

As described above, the first step of a TA project implies an appreciation of the social, political and scientific situation. According to this appreciation, the project goals will be set and the relevant methods chosen. The social, political or scientific situation, however, can change in the course of the project, so that adaptations – or at least awareness of these changes – are necessary. Thus the necessity to keep in track with the context in which the TA project is set becomes a central issue. Three ways of keeping track with social, political and scientific reality can be distinguished:

Carrying out research

Research is a good way to reflect on and keep track with the context in which the TA project is being realised. This implies the monitoring of the specialised and daily press, as well as reflection on the observed changes.

Examples: Flexibility Related to the Ongoing Debate in Practice

The TA-SWISS study on stem cells: coping with an accelerating political agenda

When TA-SWISS started its study on human stem cells in summer 2001, the Swiss Government had already decided to work out a law on research involving human subjects, in which the issue of embryonic stem cells would be addressed besides other issues. A first draft of this law was expected for the end of 2002. Knowing that this first draft would be revised after having been consulted by political parties, cantons and interested and affected groups, there was a good chance for the TA-SWISS study to be considered when writing the final law project. And of course, it might be used by members of parliament when discussing the law.

But in fall 2001, the Swiss National Science Foundation approved funding for a research project, the aim of which is to study the use of human embryonic stem cells for the regeneration of cardiac muscle cells. This gave high priority to the regulation of stem cell research in Switzerland, to avoid such research projects taking place in some sort of "legal grey area". The Swiss Government decided to draw up a separate bill on these issues in a fast-track procedure. The deadlines were shortened and it appeared that the consultation of the first draft of the law on embryo and stem cell research would take place before the TA-SWISS study was published – a problem when knowing that this consultation phase is an important step in the Swiss political decision-making system, and where TA-SWISS could have had some impact.

The TA-SWISS steering committee thus decided to accelerate the work on the TA study on human stem cells and to publish mid-term results, even though the project was not completed and did not go through all phases of quality control. (For example, the time schedule did not allow to submit the draft report for peer review). This procedure not only implied an infringement of TA-SWISS internal procedures, but also flexibility from the side of the project management and the authors of the study, as the mid-term results all of a sudden gained political relevance (Hüsing et al. 2003).

Competing for political and media attention: The Sustainable Water Management Project of the Rathenau Institute

The Rathenau Institute's project on "Sustainable Water Management" ran parallel (in time) with activities of several committees, most notably the Committee of Water Management in the 21st Century (WM21). The Dutch cabinet as a result of the flooding problems over the past decade had set up this committee during the course of the project. The 'competition' for (media and political) attention with other committees can make it hard to have an impact and to measure the effects of a project. In the case of the Sustainable Water Management project the project managers worked very hard – some even cancelled their vacation – to publish the results (in a report titled "Cashing in on the Blue Gold") on the night before the WM21 brought out its advice. If the results had been published later the WM21 advice would have probably taken away all media attention. Now the Rathenau Institute received at least as much media attention as the WM21 (Van Rooy and Sterrenberg 2002).

Meeting with experts and stakeholders

Regular meetings with representatives and experts to present and discuss project design and outcomes are important for two reasons. First, these "opinion leaders" can act as a kind of "support" for legitimization and the valorisation of the project results. Second, these persons are gatekeepers, who can be helpful to make contact with the "work floor" (the target audience).

Meeting the "work floor"

Meeting the work floor can be tough and confronting, but is a very good way for continuously adapting the project to reality. This should be done at every important stage of the project. Formal or informal interviews or organising discussion sessions can be used for this. To organise these things often the consent of the above-mentioned gatekeepers is needed.

It should be noted that the above three activities cannot always be clearly distinguished. For example, there is a strong link between interviewing a manager for a research project and asking for comments on a project design. In reality, both activities are often combined. It is important to realise that the "respondent" is not only a source of knowledge but also a kind of 'project consultant'.

Political embedding

Presenting interesting outcomes of a TA-project is often not sufficient if one seeks to create impact in political decision making process. Many TA-scientists try to

---

Examples: Keeping Track with Social, Political and Scientific Reality in Practice

The "Midway conference" for meeting with experts and stakeholders

A midway conference is a communicative method, which is often used as part of the scientific "expert discussion" at the Danish Board of Technology. This conference is arranged to discuss the very first draft of project results. Invited are a broad number of experts and stakeholders related to the topic. The midway conference includes presentations from opponents and is formed as a dialogue between the participants and the expert group responsible for the draft. Even though the expert group has already produced their first draft, the input from the midway conference often leads to the necessity for rather comprehensive editing. Therefore the name.

The "Accompanying Group": an instrument that gives further quality control

For each of its projects, TA-SWISS constitutes an accompanying group in charge of reviewing the project from beginning to the end. In case of an interdisciplinary project (i.e. a "TA Study"), the accompanying group is in charge of controlling its scientific quality: it will meet for several times in order to discuss the research questions of the project, the intermediary results, as well as the conclusions and recommendations made by the authors of the study. Accompanying groups are also set up in cases of participatory projects. In this case, they will mainly be in charge of controlling the interactive quality criteria.

Besides this primary role of quality control, the accompanying group is also a decisive element for TA-SWISS to communicate with the outside world and keep in track with reality. As a matter of fact, accompanying groups are formed of scientists from various disciplines, as well as from interest representatives. These personalities can give valuable information about scientific and technological developments related to the subject or about societal and political issues. They can also act as "multipliers" towards decision makers and other relevant actors (TA-SWISS 2003).

make contributions to the societal and political agenda setting. Unfortunately these attempts are quite often unsatisfactory and political decision making processes seem to be unaffected by TA-project outcomes. Thus part of the value of a TA-project can be seen as the extent to which the project outcomes lead to political reality. In order to improve the likelihood of reaching this objective the whole process of a TA-project (situation appreciation, goal setting, concept and design, realisation) should be embedded in the political decision making process. TA can contribute to this decision making process by offering helpful information to the decision makers, for example value balanced advice or arguments discovered by competent assembled and fairly grouped discourses. The embedding requires permanent information of the political decision makers and the societal environment surrounding them. A flexible project design is helpful in order to integrate important changes of the appraised political and societal situation in further project realisation. Active public relations work also might be helpful to some extent.

Example: political embedding in practice

The Publiforum on transplant medicine: a joint venture

The PubliForum on transplant medicine, organised by TA-SWISS, was co-financed by 2 other institutions: the Ministry of Health and the Swiss National Foundation for research. The two partners contributed to the project planning. The Ministry of Science was, at that time, writing a law on transplant medicine and the Swiss national foundation for research had launched a research programme on the issue of transplants and implants. Both institutions were interested in integrating the views of the citizen in their work. This was especially the case for the Ministry of Health, who reported on the results of the PubliForum in the message accompanying the law it addressed to Parliament (Bütschi and Mosimann 2001).

Diffusion of results

In order to play a role in social reality, convincing policy makers using the results (for instance: in writing policy advises) is not enough for a TA-project. Change is only possible if the 'policy network' or 'the public' sees the need for it and wants it. This leads to the following suggestions with respect to communication:

Publication of easy accessible media

Scientific reports are relevant for scientists and for knowledge keeping purposes within the TA institute. For addressing target groups other than scientists and the TA community, one needs to make understandable, attractive media that is easily accessible.

Communication strategy with all kinds of media

Good media communication is very important. Not only with scientific journals or quality newspapers, but especially with ordinary, popular newspapers and journals, as well as with radio and television and especially internet media.

Informal and continuous communication

A TA-practitioner's job is not finished when the project is completed and the results are published. This is when the real work begins. In this respect, in order to have some impact, it is important to talk with people, and discuss the results with them. Not one time, but over and over and in a way that people can understand. In other words, TA is also happening out of the office and out of the meeting rooms!

Examples: Diffusion of Results in Practice

Combination of Press conference and workshop
The Europäische Akademie GmbH presents the reports of their project groups in a two step process that fits the broad group of addressees the Academy is aiming at. The presentation starts off with a press conference and continues with a short workshop that allows a closer look at the report and a more intense discussion, particularly of the policy recommendations. In order to allow for a participation of the relevant decision makers and interest groups the presentations are held in Brussels and/or Berlin respectively.

Pork Plaza: a video movie to foster debate on the future of pig farms
To foster a broader debate – i.e. also involving non-specialists – about the future of Dutch husbandry, in particular pig farming, the Rathenau Institute produced in 2002 the video movie Pork Plaza. In the movie two future scenarios are shown: organic farming and the so-called pig flat, in which the whole production process (from seed to stake) is organised within a factory which has multiple levels. This video was and still is being used for discussion within secondary schools, universities and interest groups. The video has also been used by specialists, for example, within the Ministry of Agriculture.

CD Roms as part of the viWTA project on GMO-food
The viWTA-project on GMO-food consisted of three main activities: (1) a preliminary study describing the stakeholders, their positions and the ongoing debate on GMO's in Flanders, (2) a consensus conference and (3) a stakeholders forum. For the invitation for the stakeholder forum, a CD-Rom with animated photo coverage of the consensus conference was made. A voice-over described the whole process, seen from the point of view of the members of the lay panel. All related documents (report of the lay panel, information brochure, etc.) were consultable on the CD-Rom. The response to the invitation was very high and there were a lot of positive comments on the user friendly way we the results were presented. (http://www.viwta.be/content/nl/inf_publieksforum.cfm?favlang=nl)

Conferences as a way to disseminate results of TA projects
TA-SWISS regularly organises one-day conferences dedicated to disseminate and discuss the results of its studies. These conferences are aimed at experts, decision makers and/or stakeholders. In order to enhance the visibility of the conference and to foster commitments from the participants, TA-SWISS favours joint-ventures with relevant actors and institutions. For example, TA-SWISS recently organised a conference on the chances and risks of telematics in the area of transports ("Chancen und Risiken der Verkehrstelematik") after having published a study on the issue. This was an opportunity to discuss the results of the study with transport experts and stakeholders. TA-SWISS also organised a seminar with doctors and nurses intended to discuss the recommendations the citizens made during the PubliForum on transplant medicine. Participants really appreciated the seminar, as nurses and doctors rarely have common conferences to discuss societal issues related to transplantation medicine.

Attractive products and communication tools

In order reach a wider audience, it might be useful to develop attractive products and communications tools, such as personal presentations, (simulation) games, theatre, movies, interesting slides or DVD-presentations, events and expositions.

Striving for synergies

As noted before, TA projects can be conceptualised as a combination of scientific, interactive, and communicative methods. These three classes of activities supplement each other. It is very important to look for synergy between them. From a communication point of view it is also interesting to co-operate with other organisations, which bring other types of experiences and relations with other types of networks.

* *Combining science (content) and communication*

    There are various ways in which science and communication can strengthen each other. For example, adding an opinion poll to the public panel may strengthen both the results of the public panel and create interesting, ready material for the press. Organising a scientific conference can well be combined

---

Examples: Striving for Synergies in Practice

World Exhibition, Conference, and Television Show

During the World Exhibition in 2000 in Hanover, Germany, a series of conferences, open to the visitors of the Expo were organised under the heading Global Dialogue. From July 11–13 the theme was Science and Technology – Thinking the Future. At the end of each conference day, a television show was broadcast from the Expo. This show discussed themes similar to those discussed during the expert workshops, and invited distinguished scientists that visited the conferences to take part.

Conference and Exhibition

The Akademie für Technikfolgenabschätzung in Baden-Württemberg combined its conference Zur Zukunft des Menschen: Gentechnologie, Nanotechnologie, Künstliche Intelligenz with the exhibition Erde 2.0, which was targeted at a large public.

Science Museum and TA-Institute

In co-operation with the science museum Nemo in Amsterdam the Rathenau Institute organised on November 1, 2003 the technology festival Homo Sapiens 2.0. This was a large public festival on the possibilities and risks of all kind of existing and future technologies that might 'improve' men's physical, mental or abilities, looks, et cetera.

Theatre as an input for scenarios discussion

The viWTA plans for 2004 a project on 'the future of elderly in the technology society'. Threesome scenarios on the outlook of the information society within 20 years and the way elderly will deal with all these changes will be developed. Once the scenarios are written, a theatre company will be asked to make an adaptation of the scenarios for stage and several performances will be organised for a specially invited public of older people. After the performances, a discussion with the public will be initiated in order to find out what their favourite scenario would be. The analysis of these discussions can form the basis for a back casting exercise.

with making a TV program. The TV program could use the same experts, but would question those experts from a different perspective. Another interesting option would be to usemedia attractive forms like literature (collection of short stories), film or theatre (*Pig in the Middle*). Finally, the presentation of project results or debates could be linked to other events.

- *Co-operating with other organisations*

  An effective way to bring in knowledge and social connections is to co-operate within TA projects with other organisations. It can be very helpful to (temporarily) attach people from other organisations to the project team of the TA institute. As a matter of fact, the cooperation institute has its own network and addressees and can thus multiply the diffusion channels. Here, however, special attention must be given to the partners, who should fulfil the same conditions as TA institutes, i.e. credibility and independency.

1.8
Project implementation

The project implementation phase is the realisation of the project design. The distinction is sensible because it enables the identification of discrepancies between the ideal case (in mind) and the real case., The quality of a TA-project can be ensured by both the adequate choice of methods and their realisation. For example, if the decision is that a citizen jury would be the best method to reach a certain goal, one refers to the ideal case "citizen jury" in this decision process. The realisation might be deficient in two ways. Firstly, it could happen that the goal (e.g. reaching a decision by the jury) could not be reached due to conflicts within the jury that cannot be resolved. Secondly, it could happen that the project management was not able to combine an unbiased jury. In the first case the deficit would be referring to the original goal: The citizen jury had been chosen in order to reach a certain goal within the whole TA-project design. This goal was not reached. In the second case a deficit according to the above mentioned more general quality criteria occurred. On first sight the goal has been reached, i.e. the citizen's jury came to a common decision. But on second sight it was realized that the jury was biased. The project management failed to combine an unbiased jury and therefore was not able to meet the general quality criterion (implicitly co-notated to the method "citizen's jury") "neutrality".

The general quality criteria mentioned in chapter 6 (selection of methods) are of crucial relevance in the implementation/realisation phase, too. During the design phase these quality criteria are referred to, by selecting methods *in order to* achieve them e.g. scientific reliability or social fairness. During the project implementation phase these quality criteria *are used* to evaluate the ongoing process. The project management will be confronted with the need to re-design the project in some cases. The project design must consider cases like these in a flexible way:

- with reference to the individual goals each selected method within the project design should reach;

- with reference to the general quality criteria, if deficits occur during the ongoing process;
- with reference to the above mentioned "keeping contact" to the ongoing social debate. It might be necessary to adapt the project design to changes identified.

1.9
Summary and conclusions

The TAMI method group aims, at a very general level, at improving Technology Assessment (TA) and its use for policymakers by looking at the methods to be applied and at the relationship between method and impact during the many phases of a TA project. The results are based on discussions and thinking that rely on the TA experience of the institutions and persons involved. Technology Assessment is regarded by the TAMI network as a scientific, interactive and communicative process with the aim to contribute to public and political opinion forming on science and technology related societal aspects (cp. the TA definition and its explanation in part 1). Experience in TA practice has shown that there is no one single and best solution for doing TA projects. On the contrary, due to the large variety of problems to be tackled and situations to be taken into account, different types and mixes of methods have to be used. The success and impact of a TA project decisively depends on the appropriateness of the method mix used. The TAMI method group has focussed its attention on the dependency of the success of a singular TA project on many phases (situation appreciation, goal setting, project design etc.) and on quality criteria in order to provide knowledge on how to further the choice of adequate methods in those phases.

1.9.1
Functions of methods for reaching TA goals

TA projects will achieve certain goals that are dependent on the situation and the problem to be dealt with (cp. part 4). Methods to be applied are part of the design of the TA project. They are regarded as the means which serve certain functions in the TA project and which shall contribute to the success of the respective project. The most important functions of methods are, due to the several aspects mentioned in the TA definition:

- *Creating and providing relevant TA knowledge* (on impact and consequences of science and technology, on material and energy flows, on societal framework conditions, on actor constellations etc.). This knowledge can be used in the respective TA, for example, for early warning against risk, for early detection of chances, for identifying obstacles to innovation processes, for creating awareness, for supporting societal learning processes or feedback loops, for motivating and informing public involvement etc.
- *Involving stakeholders and people affected in the TA process* (for early detection or prevention of possible conflicts, for raising awareness, for enabling social learning processes etc.).
- *Motivating and structuring interaction and communication* (for supporting decision-making and opinion-forming, for analysing the value dimension, for identi-

fying alternative solutions to the problem under consideration, for testing ways of conflict resolution etc.).
- *Ensuring transparency, reliability and validity of the results* (to open the TA results for criticism, to allow transparent discussion, to allow everyone to prove in which way and under which premises and circumstances the results have been produced etc.).
- *Communication of the results* to the specific customers (like parliaments) or to the broader public.

In the specific context of a TA project these different functions will have different importance, and there will be different functions the method mix selected should serve.

1.9.2
The method toolbox

The question arises which methods are we talking about? Which methods are included in the TA toolbox? The answer can be given according to past experiences with certain methods. It cannot, however, be concluded that there will be a fixed boundary to the TA toolbox. New challenges for TA might cause the need for new methods to be used. In TA history, the TA toolbox has been increased by more and more elements, and new types of methods have enriched the toolbox. A first wave of changes supplemented the *scientific methods* with *interactive methods (like participatory or dialogue methods)*. They complement each other well and are now considered as current TA practice. Furthermore, the TA community has become aware of the importance of communicating to the outside world about the TA approach and methods, process and output over the last few years, so that *communication methods* are under development now. They complement scientific and interactive methods in current TA practice, so that the methods used in TA projects can be seen as a combination of scientific, interactive, and communication methods. According to the institutional setting of the TA, the problem to be dealt with and the respective situation parameters, this mix might include all three categories or only one or two of them. Also the relative importance of the three categories will vary from case to case. In order to illustrate the content and the composition of the TA toolbox, we will mention the most important (mostly used or of high importance in the TA discussion) elements briefly.

Scientific methods: Usually these have been developed in disciplines of natural or social sciences and then have been applied to TA problems. These methods are dedicated to the collection and evaluation of data of any kind, to allow prediction, to make quantitative risk assessments, to allow for the identification of economic consequences, to investigate social values or acceptance problems, to enable for eco-balancing etc. Methods of this type are the Delphi method, expert interviews for collecting expert knowledge, modelling and simulation, cost/benefit-analysis, systems analysis, risk analysis, material flow analysis, trend extrapolation, scenario technique for creating knowledge to think about the future (currently often combined with participatory elements in technology foresight exercises, TATuP 2003), discourse analysis, value research, ethical analyses, value tree analysis.

Interactive methods: These organise social interaction in order to make conflict management easier, to allow for conflict prevention, detection or resolution, to bring together scientific expertise and citizens, to involve stakeholders in decision-making processes, to mobilise citizens for shaping society's future or to mobilise "local" knowledge. To this method type consensus conferences, co-operative discourses, public expert hearings, focus groups, citizens' juries belong, currently in part supported by using electronic media. There may also be different mechanisms for recruiting the participants.

Communication methods: These are used to communicate the corporate image of a TA institute, the TA approach, the TA process and product to the outside world (policymakers, stakeholders, the general public) to increase the impact of TA. Communication methods also help keeping track with an eventually varying socio-political environment. Newsletters, opinion articles, science theatre, (interactive) websites and various types of networking are belonging to this class.

1.9.3
How to select the appropriate TA method

A main step in designing a TA project is to select the methods to be used and to clarify their interfaces and their integration in a coherent method mix. The question of how to optimise this step in order to arrive at the desired impact is at the heart of the TAMI programme. The objective for the method composition process is the development of a TA-design, which can be justified as the one with the highest potential for reaching the identified goals of the entire process. If an error or a misconception occurs at this point of a TA project there will be few chances to modify the method mix at a later stage of the project without a considerable loss of resources and time. However, in order to maintain and to increase the chances to adapt the respective project to changing socio-political framework conditions or situations it is sensible to implement entry-points for a flexible response to such changes.

TA activities are always located in a specific context defined by the issue to be tackled, the institutions involved, the societal and political climate, the need for opinion-forming and decision-making, the constellation of the relevant actors, the potential for social conflict, just to cite some examples. Also, the way a TA project is implemented and communicated may play a decisive role in the impact that it has.

Following previous chapters, a TA issue can lead to quite different questioning in different contexts. The way to tackle an issue might depend, for example, on the technological development, on the political context or on the general mandate of the TA institution. An adequate appreciation of the situation is therefore an important step towards having any impact. The right appreciation of an issue and its context will help a TA institution to fix realistic and correct goals for a given project, and then to use the appropriate methods in order to set up a TA approach able to realise these goals. A situation appreciation consists of analysing several aspects which depend on the respective case. Relevant to TA are – mostly – the following aspects:

– knowledge aspects: What knowledge (about impacts and consequences of technology, about reactions of stakeholders and the public) is already available? Where are important areas of uncertainty and ignorance?

- thematic aspects: which TA relevant properties does the topic under consideration show? What kind of problem (technological or scientific development,
  social problem, decisions to be made) has to be addressed?
- actor aspects: which actors are already involved? How are they organised and
  what can be said about their interrelationships? Which role does the public play?

One of the main findings of the TAMI method group is to put emphasis to the
importance of this situation appreciation. In an ideal case, there are two types of situation appreciation. At first, it should be a separate stage in the pre-course of TA
projects, because of its relevance to their later success but also in order to make the
design and the premises and presupposition as transparent as possible. Secondly, it
is highly important to constantly monitor the societal situation during the course of
the project in order to keep track with possible modifications of the initial situation
appreciation – in order to enable adequate and early reactions.

1.9.4
Quality criteria

In implementing the methods TA practitioners have to be aware of the necessity to
fulfil a certain set of quality criteria. These have been ordered in TAMI following
the above-mentioned types of methods:

Scientific quality or *reliability* of TA knowledge and orientation often seems to be a necessary but not sufficient prerequisite for a "good" TA. Bad or invalid knowledge will be
uncovered as being bad or invalid, and the consequence will be a loss of credibility and of
impact. Scientific TA as interdisciplinary research is confronted with two categories concerning the reliability of scientific statements: the reliability of the respective disciplinary
inputs (according to the familiar disciplinary quality criteria) and the reliability of the
integration of the interdisciplinary results (composed out of the disciplinary inputs).

Interactive quality criteria are related to the design of participatory or dialogue TA-
processes. They are, according to TAMI discussions, social fairness *(fairness of the
project structure, fairness during the process)*, argumentative quality, plurality,
transparency as well as support towards the competence of the participants. They
are essential in ensuring and demonstrating the legitimacy of the TA process which
is essential for the acceptance of the results in the outside world (Grunwald 2000).
Prominent aspects of realising these criteria are approaches to identify or select the
participants, and how to define the rules of the process.

Communicative quality criteria are the flexibility related to the ongoing debate, the
necessity to keep track with social reality, the diffusion of the TA results and striving for synergies. Their observance will guarantee that there is a feedback between
the TA practitioners and participants on the one hand and the outside world on the
other, in order to avoid a mismatch between external expectations and the results
provided, and in order to prepare the addressees for the coming results and to
increase the external resonance (and impact).

TAMI emphasises the importance of quality assurance in TA in order to gain
trust and credibility in the long run. The three types of quality criteria together

define the professional standards and ethics of the current TA community. These standards have been developed over time. For example, one might say that in the 1970s emphasis was laid on scientific quality criteria. Fulfilling those criteria was expected to be sufficient in order to generate an impact. After interactive quality criteria had gained weight during the 1990s, TAMI introduces a third set of communication quality criteria to indicate that communication activities and the question of how to increase impact are at the centre of attention of current TA organisations. Having an impact is important but TA should never go for easy publicity. Greater impact may be achieved through strengthening communication competencies and activities, but not at the cost of scientific and interactive quality criteria. As such the three types of quality criteria together enable and constrain short and long-term impact, and the way in which these impacts should be reached. Which criteria are most relevant in specific exercises depends on the concrete challenge, the goals and the respective situation.

1.9.5
Recommendations concerning the relationship between method and impact

In this paper we have set up a common framework to get an insight into the complex relationship between method and impact. Elaborating on this framework leads us to the following list of recommendations.

- Set proper and realistic project goals and choose the appropriate (mix of) methods based on a sound and detailed situation appreciation.
- Be aware that there is no unique answer to a situation appreciation, in other words, the same issue can be addressed in different ways. For example, depending on its institutional context, one institute might address the lack of knowledge about an issue, whereas the other might focus on stimulating public debate.
- A TA institution should be aware of its way of tackling issues and that it is known and accepted by outside players (e.g. decision-makers). Only under these conditions will it be possible to set the right goals and to define an appropriate project, with chances to reach a certain audience – and thus a certain impact.
- Be aware that gaining any type of impact is not an end in itself. Instead of solving a problem, a TA project could create new problems. In other words, negative impacts may occur. It is important to be conscious of the appropriate type of impacts one wants to strive for.
- Be aware of the fact that no simple linear relationship exists between the methods used and the impact achieved. The impact of TA projects also depends on both internal factors (like project management competencies, budget constraints, organisation culture, and institutional arrangements) as well as (mostly uncontrollable) external influencing factors (such as strategies of other actors, timing of the policy-making process, sudden changes in the problem situation). Accordingly, a proper choice of methods does not guarantee reaching the project goals.
- Ensure that the TA project keeps track with social, political and scientific reality. Situation appreciation should, therefore, be a constant feature of project management.

- Allow for sufficient flexibility in the project design and the procedures of the TA institution to adapt to relevant changes – like new scientific discoveries, an increasing (and often sudden and short) media attention, a political intervention, etc. – that may (and most likely do) occur during the duration of the project.
- TA projects have to fulfil scientific, interactive and communicative quality criteria in order to have legitimate short-term impact and to build up and maintain institutional trust in the long term.
- Be aware that there might be possible ambiguities or trade-offs between short-range impact and the building of long-term trust and credibility. Creating impact by using invalid or deficient knowledge is, for example, no problem for mass media; their aim is to have a short-term impact and create awareness. TA, however, would endanger its long-term credibility by using such an approach.
- One should, therefore, distinguish between impacts on different time scales.
- Consider communication as the key manner to achieving impact. Fulfilling scientific and interactive criteria is a necessary, but not sufficient, condition for having an impact.
- Realize that communication goes beyond communicating about the results of a project. Selecting a certain topic for the agenda of your institute already signals a message to the outside world.
- Realize that you are doing TA and assessing societal issues related to science and technology is your core business. Strengthen this TA identity by developing communication competencies in-house, but do not become a communication bureau. Look for synergies with organizations – like science museums, debating centres, and media – that are specialized in communicating messages.

References

Biesboer F et al. (1999) Clones and cloning: the Dutch debate. The Hague: Rathenau Institute; Working document 70

Bröchler S, Simonis G, Sundermann K (eds) (1999) Handbuch Technikfolgenabschätzung. Edition Sigma, Berlin

Bütschi D, Mosimann F (2001) Médecine de transplantation: un débat de société, In: Revue médicale de la suisse romande, 121, pp 91–94

Carius R, Renn O (2003) Partizipative Risikokommunikation. Wege zu einer risikomündigen Gesellschaft. Bundesgesundheitsblatt. Springer, Berlin

Coenen R, Grunwald A (2003) Nachhaltigkeitsprobleme in Deutschland. Analyse und Lösungsstrategien. Edition Sigma, Berlin

Decker M (ed) (2001) Interdisciplinarity in Technology Assessment. Implementation and its chances and limits. Springer, Berlin

Decker M, Grunwald A (2001) Rational Technology Assessment as Interdisciplinary Research. In: Decker (2001), pp 33–67

Decker M, Neumann-Held E (2003) Between Expert TA and Expert Dilemma – A Plea for Expertise. In: Bechmann G, Hronsky I (eds) Expertise and its Interfaces. The tense relationship of Science and Politics. Edition sigma, Berlin

Dienel P C (1989) "Contributing to Social Decision Methodology: Citizen Reports on Technological Projects". In: Vlek C, Cvetkovich G (eds) Social Decision Methodology for Technological Projects. Kluwer Academic Press: Dordrecht pp 133–150

Fiorino D J (1990) "Citizen Participation and Environmental Risk: A Survey of Institutional Mechanisms", Science, Technology, and Human Values, 15, No.2, Spring, 226–243

Fischhoff B (1996) "Public Values in Risk Research". In: Annals of the American Academy of Political and Social Science, Special Issue, Kunreuther H, Slovic P (eds) Challenges in Risk Assessment and Risk Management. Sage: Thousand Oaks, pp 75–84

Freeman JB (1991) Dialectics and the Macrostructure of Arguments. Foris, Dordrecht

Gethmann C F, Sander T (1999) Rechtfertigungsdiskurse. In: Grunwald A, Saupe S (ed) Ethik in der echnikgestaltung. Praktische Relevanz und Legitimation. Springer, Berlin

Goorden L, van Gelder S et al. (2003) Genetisch gewijzigd voedsel in Vlaanderen. Retrospectieve trendanalyse van het maatschappelijk debat; Brussels, viWTA rapport nr. 1

Gram S (1998) Urban Traffic – a wish for political coordination. The Danish Board of Technology, Copenhagen

Grin J, van de Graaf H, Hoppe R (1997) Technology Assessment through interaction: A guide. The Hague: Rathenau Institute; Working document 57

Grin J, Grunwald A (eds) (2000) Vision Assessment: Shaping Technology in the 21st Century Society. Towards a Repertoire for Technology Assessment. Springer, Berlin

Grunwald A (ed) (1999) Rationale Technikfolgenbeurteilung. Konzepte und methodische Grundlagen. Springer, Berlin

Grunwald A (2002) Technikfolgenabschätzung – Eine Einführung. Edition Sigma, Berlin

Grunwald A (2000) Technik für die Gesellschaft von morgen. Möglichkeiten und Grenzen gesellschaftlicher Technikgestaltung. Campus Verlag, Frankfurt/New York

Harremoës et al. (2001) Late lessons from early warnings: the precautionary principle 1896–2000. Copenhagen: European Environment Agency (EEA). Environmental issue report No. 22

Hüsing B, Engels E M, Frietsch R, Gaisser S, Menrad K, Rubin-Lucht B, Schweizer R (2003) Menschliche Stammzellen. Bern: TA-SWISS, Report TA 44/2003

Joss S, Bellucci S (eds) (2002), Participatory Technology Assessment – European Perspectives. Centre for the Study of Democracy (CSD) at University of Westminster in association with TA Swiss, London

Joss S, Brownlea A (1998) Verfahrensgerechtigkeit in der partizipativen Technikfolgenabschätzung am Beispiel des Publiforum Strom und Gesellschaft. Konzepterarbeitung und Evaluation, TA-SWISS, TA-DT 22/1998

Keeney RL (1996) "The Role of Values in Risk Management". In: Annals of the American Academy of Political and Social Science, Special Issue

Kunreuther H, Slovic P (eds) Challenges in Risk Assessment and Risk Management, Sage: Thousand Oaks, pp 126–134

Nennen HU, Garbe D (1996) Das Expertendilemma: Zur Rolle wissenschaftlicher Gutachter in der öffentlichen Meinungsbildung. Springer, Berlin

Paschen H, Vig N (eds) (1999) Parliaments and Technology Assessment. The Development of Technology Assessment in Europe. State University of New York Press, Albany

Renn O, Webler T (1998) Der kooperative Diskurs – Theoretische Grundlagen, Anforderungen, Möglichkeiten. In: Renn O, Kastenholz H, Schild P, Wilhelm U (eds) Abfallpolitik im kooperativen Diskurs. Vdf ETH Zürich

Renn O, Webler T (1994) Konfliktbewältigung durch Kooperation in der Umweltpolitik. Theoretische Grundlagen und Handlungsvorschläge. In: OIKOS, Umweltökonomische Studenteninitiative an der HSG (ed) Kooperationen für die Umwelt. Im Dialog zum Handeln, Rüegger, Zürich

Renn O, Webler T, Rakel H, Dienel P C, Johnson B (1993) Public Participation in Decision Making: A Three-Step-Procedure. Policy Sciences, 26, 189–214

Renn O, Webler T, Wiedemann P (eds) (1995) Fairness and Competence in Citizen Participation. Kluwer, Dordrecht

Renn O, Schrimpf M, Büttner T h, Carius R, Köberle S, Oppermann B, Schneider E, Zöller K (1999) Abfallwirtschaft 2005. Nomos, Baden-Baden

Ropohl G (1979/1999) Allgemeine Technologie. Eine Systemtheorie der Technik. Hanser, München. Older version: Eine Systemtheorie der Technik. Suhrkamp, Frankfurt

Slocum N, Beyne C, Steyaert S (eds) (2003) Participatory Methods Toolkit. A practitioner's manual. Brussels, viWTA – KBS

Steyaert S, Weyns W (eds) (2003) Public Forum. New impulses for the debate on genetically modified food (final report of the public panel). Brussels, viWTA rapport nr. 2, p 42

TA-SWISS (2003) TA-SWISS Portät. Akteure und Abläufe bei Projekten des Zentrum für Technologiefolgen-Abschätzung, Bern, TA-SWISS, TA-DT 30/2003

TATuP (2003) Special Issue 1/2003 of "Technikfolgenabschätzung – Theorie und Praxis" on Technology Foresight, edited by Knud Böhle and Michael Rader

Toulmin S (1958) The uses of Argument. Cambridge University Press, Cambridge

Vandenabeele J, Goorden L (2001) Leken en experten uitgedaagd? Evaluatie van door VIB georganiseerde debatavonden over biotechnologie in landbouw en voeding. Antwerpen, Zwijnaarde, UA – VIB, p 52

Vandenabeele J, Goorden L (2002) Biotechnologie en het debat anno 2002. Een vooruitblik. Antwerpen, Zwijnaarde, UA – VIB, p 55

Van Est R, Hanssen L, Crapels O (eds) (2003) Genes for your food – Food for your genes. Societal issues and dilemmas in food genomics. The Hague: Rathenau Institute, Working document 92

Van Est R et al. (2002) "The Netherlands: Seeking to involve wider publics in Technology Assessment". In: Joss S, Bellucci S (eds) Participatory Technology Assessment: European perspectives. Centre for the Study of Democracy, London

Van Rooy P, Sterrenberg L (2002) Het blauwe goud verzilveren. Aanzet voor kennisfusie voor intergraal waterbeheer. The Hague: Rathenau Institute, internal report

Wagner-Döbler R (1989) Das Dilemma der Technikkontrolle. Edition Sigma, Berlin

Webler Th, Levine D, Rakel, H, Renn O (1991) The Group Delphi: A Novel Attempt at Reducing Uncertainty, Technological Forecasting and Social Change, 39, No. 3, pp 253–263

Weisbord MR, Janoff S (1995) Future Search and action guide to finding common ground in organizations and communities. Berrett-Koehler Publishers, San Francisco

2 Towards a Framework for Assessing the Impact of Technology Assessment

Leonhard Hennen, Sergio Bellucci, Robby Berloznik, David Cope, Laura Cruz-Castro, Theodoros Karapiperis[6], Miltos Ladikas, Lars Klüver, Sanz-Menéndez, Jan Staman, Susanne Stephan, Tomasz Szapiro

2.1
Introduction

Any discussion about the mission of TA and its relationship to policy making will soon enough touch upon the question of the effects that TA might or should have on decision making as well as on the content and path of political and social debate on technology issues. TA, as an endeavour that is meant to explore the possible impact of technology on society in order to support policy making, will naturally be asked to bear witness on its own impact as an indicator as whether or not it really fulfils its ascribed mission and tasks. Discussions on impact of TA, however, usually suffer from a lack of common understanding of TA's objectives and of what can be expected as an impact of TA. Unsurprisingly, there is also very little available information on attempts to measure or evaluate the impact of TA in Europe. This reflects the past lack of coordinating action in discussing the goals of TA and the processes by which they can be attained. The knowledge vacuum in this area could influence the ability of TA as a discipline to communicate its roles and consequently, its value in society. It was the objective of the TAMI impact group to develop a structured discussion on the objectives, functions and effects of TA and prepare the ground for future attempts to evaluate TA procedures and their impact on related decision making processes. Being aware of the complex nature of the issue of impact evaluation, the group did not aim at developing a detailed scheme or a set of criteria for future evaluation procedures. Instead, TAMI aimed to provide a frame of reference on the relationship of objectives, methods and impacts of TA. The discussion among the group members – the outcome of which is presented in this paper – evolved around the question of which kind of impact can realistically be expected from TA as a particular branch of policy consulting taking into account its mission and methods, the nature of the issues it is dealing with and the characteristics of the field of policy making it is acting in. The result of this process of self-reflection among a group of TA experts from many European countries is mainly a matrix of TA impacts that helps to clarify the role of TA in technology policy and the related impacts that can be expected.

---

6 "The views expressed are those of the author and do not necessarily reflect those of the European Parliament".

2.2
The application of knowledge in policy making

The question of the impact of TA can be regarded as a sub-case of the general question of impact of scientific advice on policy making. The conceptualisation of the relationship between scientific knowledge and policy making as discussed among scholars of policy analysis during the last decades, has undergone a fundamental change. It can be characterised as a constant dissolution of what could be called a "rationalistic" concept of scientific knowledge. Central to this rationalistic concept – which is also often implicitly or explicitly referred to in discussions on the role of TA in policy making – is the notion that science provides in depth analysis of the problem at stake and explores viable ways (if not the one best solution) to deal with the problem. The results of scientific analysis are evaluated by policy makers with regard to their costs and benefits and finally are implemented in political programmes or legislation. In its strong technocratic version this concept implied the "dream of the abolition of politics" by scientific knowledge, rational policy making would simply have to follow the solutions provided for by science. This "first face" of policy analysis obviously neglected the fact that the definition of the problem at stake as well as the search for solutions is dependent on values and interests involved and that neither can science provide a one best solution nor will policy makers act according to the rational choice model of decision making (Torgersen 1986).

In recent decades, social studies of science and technology have shown that – particularly in the field of technology – scientific knowledge is necessarily incomplete, provisional and undetermined with regard to the complexity of the problems of policy making. Ethical questions growing out of scientific development as well as the assessment of risks for human health and environment cannot be reduced to scientific facts and be stripped of the values and interests that have to be taken into consideration in policy making (see e.g. Beck 1985, Functowicz and Ravetz 1992; Nowotny, Scott and Gibbons 2001). There is no clearly defined boundary between facts and values in policy consulting and normal practice in policy advice is not "speaking truth to power" but rather a negotiation about the most practicable and socially acceptable definition and solution of the problem in the light of different and contradictory scientific insights as well as the given policy frame (Jasanoff 1990).

In policy analysis a "post-positivistic" (Torgersen 1986, Héretier 1993) conceptualisation of the policy making process accommodates the fact that policy making cannot be described as rational decision making. "Garbage-can theories" – originally developed in organisational theory (Cohen, March and Olson 1972) – reject conventional "policy cycle" models which envisage policy development processes as underpinned by the logic of problem solving. Policy making is regarded as being chaotic, random and frequently non-rational. There is only a loose relationship between problems and policy solutions, since the latter is not driven by rational handling of well defined problems but bounded by given institutional structures, ideas and interests (Kingdon 1984). Post positivistic concepts stress the relevance of the complex influences of social actors engaged in policy networks on the policy making process (van Warden 1992, Scharpf 1993) and the relevance of strong "belief systems" held by competing "advocacy coalitions" (Sabatier 1988), which frame decision making and are not easily changed by input of scientific knowledge but only in long term learning processes.

Thus it would be naive to expect to track a path of insights of a TA procedure all the way down to its final impact on decisions taken. Information, data and arguments provided by TA or any other process of policy consulting cannot be regarded as "magic bullets" whose impact in legislation, regulation, budgetary decisions or the design of political programmes is directly observable. This is usually reflected in studies on the utilisation of results of TA by policy makers (Berg et al. 1978, Paschen et al. 1991). Based on interviews with TA experts and their clients (i.e. policy makers) as well as on general experiences in the field of TA the few studies available show that what can be expected is "conceptual use" (Caplan 1979) of scientific knowledge by policy makers but not "instrumental use" in the sense of using TA results directly as a "guideline" for political action. "Conceptual use" of knowledge includes: awareness of the complex interconnection of the problem under consideration with different fields of policy making, possible effects not being taken into account and change in the policy makers view on priorities for political action (Berg et al. 1978). The factors restricting the application of TA results in policy making discussed among TA experts are manifold (for an overview see Paschen et al. 1991); they include practical problems such as restricted resources, timing for dissemination of results and interaction between TA analysts and their clients, or individual factors such as lack of experience of scientific staff of a TA organisation with routines of the policy making process. A general and structural restriction however is given by the different "logics" or "rationales" of scientific analysis and political action:

– Since policy making is dominated by conflicting interests, values and beliefs, scientific knowledge has to pass through this filter and thus is modified and selected according to interests and opportunities and not applied according to scientific criteria of rationality. It is more likely for scientific knowledge to be strategically used (or not used) during processes of negotiation and bargaining according to different interests, values and beliefs.
  Scientific advice (and TA in particular) often increases complexity of decision making since it provides a full and unbiased picture of the problem (including different social perspectives and areas of uncertainty). Thus, TA cannot easily be used to foster views held by actors and almost never can provide recipes for problem solving (as often might be expected by policy makers). Instead of direct application of scientific advice, there might be long term effects on the general perception of problems and practical ways of problem solving. "Knowledge, including scientifically-produced knowledge, flows into the decision making process through obscure channels from many different sources, and this results in a more general awareness of the way the world appears and is structured." (Albaek, 1995: 85).

Besides those general structural restrictions to the application of "knowledge", there are some other problems which make it difficult to identify whether knowledge has been applied or not, or to "measure" its impact. To mention only one, single processes of policy consulting (such as a TA-project) are always only one voice in a concert of consulting processes going on[7]; this necessarily restricts the influence of individual reports and makes it almost impossible to retrace arguments processed in policy debates or the outcomes of policy making processes to one particular study. Single TA studies rarely induce totally new insights or ideas into debates and policy making, instead they strive to provide a comprehensive or multiperspective view of the problem.

2.3
Definition of impact

Given the aforementioned problems of relating the results of a TA-process directly to perceived changes in policy making or changes in public debates on technology development, it might be wise to take leave of the term "impact" in order to avoid the notion of a direct and visible influence of TA-activities on policy making and public debate. For the sake of evaluation of TA-procedures it might be more appropriate to apply more open or "soft" concepts such as "success" or "resonance" that preclude the normative and demanding connotations of "impact".

All of the three terms – success, resonance and impact – have some common connotations since they denote a kind of "effect" of a TA-process. Referring to "effects" of a TA-procedure, one might say that the process has been "successful" once the client finds the information helpful, which in the same way could be seen as an "impact" of a TA-process, and of course indicates that the process or report found "resonance" at the client's side. However, all three of them have particular connotations as well:

Success is a highly subjective concept. Whether a TA-project has been successful or not depends on the expectations of the observer. From the point of view of organisations, the survival of the organisation in a difficult environment can justifiably be regarded as "success". Of course this does not necessarily imply that the work of the organisation has had any particular "impact" on its environment. TA-studies are often appreciated by policy makers as helpful for decision-making processes. In terms of customer satisfaction, this can be seen as "success". It might however have no impact (see below) in terms of "influencing the path of decision making".

The term resonance as it was applied in a recent project on participatory TA in Europe (Hennen 2002) is an attempt to avoid the normative connotations of the term "impact" (which stems from an ideal concept of policy consulting) and apply instead a neutral concept that is suitable for the purpose of empirical description. Resonance in this sense describes any kind of observable reaction to a TA process in its societal environment. Resonance does not refer to "making a difference" in the above sense but is more generic as it includes a scope of possible "effects" ranging from e.g. "Report or TA-process being mentioned in a debate or in a newspaper" to

---

[7] For a detailed account of the effect of professional lobbies in policymaking, see "Organised Interests in S&T Policy; The influence of lobby activities" at the Supplement Papers section of this book.

"impact" in the sense of e.g. "a change in the political agenda" or "new legislation caused by TA".

The term impact refers to the expectation which, on a general level, is held by both TA-practitioners and clients (e.g. policy makers) as well as observers of policy consulting. TA has to make a difference in terms of the quality of decision making processes by adding comprehensive and non-biased knowledge to this process. The implicit expectation here is that decision making with TA leads to "better" (more rational, informed or legitimate) decisions than would have otherwise been achieved without TA. This is however based on an ideal concept of rational decision making (which to some extend ignores the reality of politics) and the impact of TA in this sense is hardly measurable. Nevertheless this concept is behind all discussions on impact since it is indeed connected with the traditional mission of TA.

For the purposes of the TAMI project which is not dedicated to gathering data and/or empirically exploring the effect of a TA procedure in its environment, but rather to furthering the discussion between TA practitioners and clients on the relationship between methods applied and impacts achieved, it was decided to use the term "impact" in the sense of "making a difference" (in a more general sense however, that does not relate it to the hardly measurable goal "improving decision making in terms of rationality or legitimacy"). Impact is understood as an effect of TA (its results and the process) on what is "known", "debated" and "done" in debates on technology policy issues. Hence,

> "Impact of TA is defined as any change with regard to the state of knowledge, opinions held and actions taken by relevant actors in the process of societal debate on technological issues".

This concept does not escape problems of measurement or visibility of impacts. Nevertheless, one might succeed in developing a platform for self-reflection among TA-practitioners and clients on the relationship of TA's mission, its methods (i.e. what TA does or can do) and the role TA might play in the context it is working in or is expected to support. The range of possible or conceivable impacts covered by this approach is in a way as broad as possible: from raising awareness for a particular issue/problem to changing legislation. It is by opening up a comprehensible tableau of impacts in the sense of "making a difference" that we hope to support the understanding of what contributions (and under which particular conditions) can realistically be expected of TA to policy making as well as public debate.

2.4
Typology of impacts

This approach has been clarified by working out a typology made up of three dimensions of impacts which can be related to three dimensions of the issues that TA is dealing with. In general, one can discern three dimensions of impact that TA or policy consulting could be expected to have: impact in the sense of *raising knowledge* on issues among policy makers or in public debate, impact in the sense of *forming opinions/attitudes* of actors involved in policy making and the debate, and impact in the sense of *initialising actions* taken by policy makers or other actors.

These dimensions of impact can be linked to the three dimensions of the issues that TA-projects usually deal with and TA is expected to generate knowledge about. TA has to deliver (as comprehensively and unbiased as possible) information on the *technological and scientific aspects* of the issue that is at stake (e.g. features of technology, results/or problems of scientific risk assessment, economic costs, eco-balances, etc.). A description of the problem or issue at stake would be incomprehensible without describing the *societal aspects*: TA has to deliver knowledge about the relevant actors (their interests, values etc.) and the possible social conflicts that can evolve around the technology under consideration. On the grounds of a proper description of the scientific and technological aspects and in connection with a description of the social environment (debate, actors), TA has to analyse the *policy aspects* of the problem; i.e. it has to consider the restrictions and opportunities of policy making and has to develop policy options, such as exploring politically viable ways for problem solving (legislation, R&D funding, action plans). At the end, it has to again evaluate policy options with regard to possible side-effects (e.g. social conflicts) they might produce.

Generally it is possible to translate the term "impact" (of TA, of policy consulting) by "application of knowledge/information". The three dimensions given above then might be read as an application-continuum leading from "raising knowledge" to "forming attitudes/opinions" and to "initialising action/initiative". In the first dimension one could speak of a low level of application in the sense of "the client taking notice of results of a TA process", which may imply the user having a "fuller understanding of the problem" or "a broader view of aspects related to the problem" without directly (or visibly) inducing a change in attitude or behaviour. This is the necessary first step to a more explicit application in the dimension of "attitude" and "action": the *application* of (new) knowledge stemming from TA studies becomes visible (in the agenda and in policies) and knowledge is observably applied as arguments in the debate. In a further step this might have visible impacts on decision making in the sense of changing the path of policy making and bringing up new political initiatives. The latter two steps obviously imply not only awareness but application of knowledge in a narrower sense and therefore presume an active adoption of knowledge by the user – it has to be integrated into the "belief system" of actors as "conceptual knowledge" (see above). The application of knowledge for guiding political action is, apart from its adaptability to the knowledge and interest structure of actors, highly dependent on constraints and opportunities given by the actual policy situation (e.g. need for compromises in policy networks, respect for existing policy coalitions, compatibility with existing policy programmes etc.).

It is therefore appropriate to differentiate the three dimensions in terms of level of impact. With regard to the influence of TA on the policy making process it is more likely (and can be regarded as a relatively minor effect on policy making) to make actors aware of possible unintended consequences of technology, perspectives of actors and policy options ("raising knowledge"), than to induce a change in the agenda or political initiatives (e.g. new legislation). The three dimensions of the typology should not however be read as a continuum in time, i.e. that first a TA project is carried out, then the results are delivered to the client and the public where they may induce a learning process which again may lead to new initiatives in policy making or to decisions that close the debate. Impacts in all three dimensions may result throughout the entire TA process. Already at the beginning of a TA process, discussions between TA-practitioners, the client, stakeholders and experts about appropriate problem definition, questions and problems that should be scrutinized, may induce a learning process and result in a change of attitudes or opinions of relevant actors. Since a TA-process usually implies communication between the TA-institute and the client throughout the whole process (and in particular when communicative methods like workshops, participatory procedures are part of the process) a change in attitude and opinions and even effects on ongoing decision making may be induced in any phase of a TA-project.

Table 2.1: Typology of impacts

Impact Dimension Issue Dimension	I. Raising Knowledge	II. Forming Attitudes / Opinions	III. Initialising Actions
Technological/ Scientific Aspects	**Scientific Assessment** a) Technical options assessed and made visible b) Comprehensive overview on consequences given	**Agenda Setting** f) Setting the agenda in the political debate g) Stimulating public debate h) Introducing visions or scenarios	**Deframing of Debate** o) New action plan or initiative further scrutinise the problem at stake p) New orientation in policies established
Societal Aspects	**Social Mapping** c) Structure of conflicts made transparent	**Mediation** i) Self-reflecting among actors j) Blockade running k) Bridge building	**New Decision making Processes** q) New ways of governance introduced r) Initiative to intensify public debate taken
Policy Aspects	**Policy Analysis** d) Policy objectives explored e) Existing policies assessed	**Re-Structuring The Policy Debate** l) Comprehensiveness in policies increased m) Policies evaluated through debate n) Democratic legitimisation perceived	**Decision Taken** s) Policy alternatives filtered t) Innovations implemented u) New legislation is passed

Using these dimensions of impact and issue we have derived a matrix that shows nine types of impacts of Technology Assessment. An inventory of 21 roles or functions of TA in policy making , developed by members of the TAMI-project on the basis of their experience as TA-practitioners as well as by referring to existing case studies on the political role of TA procedures (cf. Buetschi/Nentwich 2002), can be described according to these types of impact. For every role a short example from the practical work of European TA institutions is given in the annex of this chapter.

2.4.1
Raising knowledge

"Raising knowledge" can be seen as the "classic" mission of TA. The establishment of TA in the 1960's and its development in the following decades was to a great extent encouraged by policy makers who perceived a lack of access to reliable scientific information on modern technologies that were growing in importance for almost every field of policy making (National Academy of Sciences 1969, Hetman 1973, Vig and Paschen 2000). The perceived deficit referred to knowledge on hard scientific facts (on features of technology), the socioeconomic context relevant to the implementation of technology, social needs and interests that might cause conflicts in implementation, as well as on viable and socially accepted policy options to shape or steer technology development.

The three types of impact in the column "raising knowledge" are directly related to the content of a TA process and to the "deliverables" of TA. The outcome of a TA-process (e.g. a report) as well as the process itself (participatory procedures, workshops etc) makes policy makers or other relevant actors aware of new aspects of the problem/issue at stake. Examples of this are scientific knowledge on paths of technology development, risks, chances, unintended consequences etc. (*scientific assessment*), interests or perspectives of actors involved (*social mapping*) and problems and options of policy making (*policy analysis*).

Providing knowledge on scientific, social as well as policy aspects of technology in the sense of "making the client more aware" is closely associated with the quality of TA as a scientific process. Nevertheless it is not restricted to a particular type (i.e. classical scientific policy consulting type) or method of TA. The quality of the knowledge provided may depend on scientific standards of the TA process as well as on the level and quality of inclusion of stakeholders and societal groups. Quality of the output – no matter if this is a written (scientific) report or the results of workshop – is a prerequisite for "raising knowledge". It is not however a warrant: whether the results of a TA process are taken into consideration or not, depends to a large extent on factors such as visibility, timeliness of the process as well as on contextual factors which are out of the reach of a TA institution (see below, section on influencing factors).

2.4.1.1
Scientific assessment

Scientific assessment comprises two classic roles of TA that are related to its function of making comprehensive scientific knowledge available for decision makers.

(a) Technical options assessed and made visible

With regard to the rapid development of science and technology and the overall dependence of social welfare on the application of R&D, there is a need for policy makers to be informed about what is about to come and to compare different possible paths of technology development. Making technical options visible and assessing the viability of different technological paths by means of foresight studies or scenario writing, is a prerequisite for rational decision making in innovation policies.

(b) Comprehensive overview of consequences given

Technology foresight is in most cases not a TA task in itself. It is however a prerequisite for providing a *comprehensive overview on consequences* connected with a technology. Awareness of the fact that scientific progress is often connected with unintended consequences for society, economy and environment, was the reason for a growing demand to apply scientific methods in anticipating long term consequences. To this end TA draws on a range of scientific procedures such as risk assessment or economic modelling. The added value of TA lays in a comprehensive overview of possible effects – not only economic cost-benefit calcualtions – as a fundamental prerequisite for policy making (OECD 1983, Paschen and Petermann 1991).

2.4.1.2
Social mapping

The assessment of the pros and cons of technological innovation has to be grounded as much as possible in scientific data. However this is necessarily connected with value judgements. Controversies over technologies make it obvious that different social groups arrive at different judgements depending on their interests, preferences and values.

(c) Structure of conflict made transparent

For the decision-maker it is necessary to know about the *structure of conflicts* around the technology at stake, in order to strive for consensus or compromises and for a decision that can hopefully meet acceptance and can be regarded as legitimate. This is important since different interests and values will have to be taken into account. TA might supply this knowledge by means of social research (surveys, discourse analysis, focus groups, etc.) as well as by participatory methods: i.e. giving every relevant group a say in the process (workshops, advisory boards, etc.).

The analysis of the differing preferences, interests and values that are behind conflicting expectations and demands can expand the understanding of the social context of policy making and may provide opportunities for conflict resolution. It can also be seen as an integral part of the assessment of risks and benefits since such assessment depends on the values held by the assessor. Discourse analysis, used to clarify the interconnectedness of scientific arguments and expert judgements in debates that revolve around ethical beliefs and world views, may separate facts from values and establish awareness of the fundamentally political character of technology debates which on the surface might appear as debates on scientific facts.

2.4.1.3
Policy analysis

In general TA can be regarded as policy analysis as it aims at improving the "quality" of policy making by analysing the contextual boundaries and opportunities as well as the goal setting of policy making. With regard to technology policy this necessarily includes what is called here "scientific assessment" and "social mapping". However, raising knowledge on scientific and social aspects of an issue can be seen as precondition for the conduct of "policy analysis" in a narrower sense – i.e. the evaluation of options for policy making addressing the question: Which goals and measures are available and at what cost can they be achieved?

(d) Policy objectives explored

In case of new options for emerging technologies it is often unclear what objectives policy makers can or should go for. Additionally, there is usually dissent among relevant actors on the usefulness of applying the technology or the direction in which options for the technology should be shaped. It is one of the classic tasks of TA to explore policy objectives with regard to their viability, social acceptability, the instruments at hand and the possible side-effects that might appear (e.g. the technological field of genetic engineering opens up a broad scenario of applications). With regard to policy making it is crucial for TA to structure the field by discerning applications that are likely or unlikely to be realised within a certain period of time. In addition TA needs to explore the degree to which a wide range of social needs is met by different technological options. Based on this, TA tries to evaluate different objectives that policy makers might opt for as well as opportunities and costs (economic and social) that might be incurred to achieve these objectives (funding of research for particular applications of technology, legal regulation to exclude unintended effects).

(e) Existing policies assessed

In most cases political and social debates have already come up with a range of policy options. Those options are often based on conflicting preferences – e.g. economic vs. environmental. It is then a crucial task of TA to contribute to decision making by assessing existing policies with regard to the different preferences and assumptions on which they are based. As well as this, TA needs to explore probable effects and the effectiveness of instruments (legal regulation, voluntary agreements, financial measures like environmental tax etc.) for the different policy options. It might also be that an evaluation of different national policies with regard to a technology is necessary to benchmark policies and to deliver information on options for an internationally sound system of regulation.

2.4.2
Forming attitudes/opinions

Raising knowledge is a precondition for changing the opinions and attitudes of actors. If this is achieved, the structure of debate and the policy making process has also changed in some way. Changes in attitude may occur with regard to new scientific aspects which are now discussed among policy makers or in public debates (*agenda setting*). It may happen that the TA-process or outcome change the way that the relevant actors see each other or deal with each other (*mediation*) or that options for policy making are seen/discussed in another way or that new options become prominent on the agenda of policy making (*restructuring the policy debate*).

An impact in the dimension "attitude and opinions" implies that TA induces a kind of learning process among actors and thus affects the ongoing social debate. In this respect it is important to stress the difference regarding mission and methods between the two major TA-paradigms, since this reflects a difference in intervening in the process of opinion forming. According to the "classic" (OTA like) TA approach, the mission of the TA-institute is mainly[8] to scientifically analyse the issue at stake and deliver unbiased and as comprehensive as possibleknowledge about the technical, social and policy aspects of the issue to the client (policy maker). This usually results in a written report, which is delivered to the client at the end of the project. So it is mainly by communicating the *outcome* of the TA-project to the client and the public that the TA-institute intervenes in public discourse and thus may induce an impact in terms of agenda setting, mediation and restructuring the policy debate. TA institutes following the paradigm of "public TA" intend (and are expected) to directly intervene in or organise "the public debate". Thus the whole TA-process is regarded as an integral part of the societal debate and communicative and/or participatory procedures are not only applied as instruments to gather knowledge or communicate results but also as the core of the project and an end in itself. So "public TA" directly intervenes in the ongoing process of opinion forming that is made up by the societal debate, and public TA has developed particular methods to do so. The classical TA-process is somewhat "detached" from the debate and is only intervening indirectly by communicating the results "from outside".

2.4.2.1
Agenda setting

The knowledge on technical and scientific aspects of the issue at stake provided by a TA process can affect the agenda of public and political debate by expanding the scope of aspects taken into account or by shifting the attention of debates to a new problem.

(f) Agenda-setting in political debate

When the results of a TA-process are regarded as something that needs to be considered politically – either because results are seen as supportive for ongoing policy making processes or because they are taken up by relevant actors or have found resonance with the public (media) – TA can change the agenda of the political debate on the issue or even initialise a political debate on another issue that had not been considered as relevant to politics so far. This does not necessarily result in political decisions but at least in stated political awareness. TA institutes working on behalf of parliaments can contribute in several ways to the agenda setting processes in the

---

8   The two TA-paradigms have to be understood as "ideal-types". Most TA institutes include (with different weighting) methods of scientific policy consulting as well as methods of public TA.

parliament. In some cases, written or oral comments are asked on governmental policy papers, in other cases the TA institute is invited to give comments during a hearing of a specific parliamentary committee and in other cases a committee might ask the TA institute to provide additional information on a specific issue.

(g) Stimulating public debate

TA can be understood as a formal (organised) procedure of debate on technology that is related to the informal modus of technology assessment that is going on in societal technology controversies (Rip 1986, Hennen 1999). Even if a TA-institute's mission is restricted to policy consulting, a TA process can have an impact on public debate in the sense that a new issue is taken up by societal actors and the interest of the general public is stimulated. When TA-organisations have the mission to *stimulate debate* – they take over the task to directly set the agenda of a debate by means of organising public events.

Especially in the case of new emerging technologies with a high degree of uncertainty about effects and ambiguity on their moral acceptability, stimulating debate among organised actors (scientists, NGOs, representatives of political parties) as well as among the general public can be necessary to explore the pros and cons as well as possible areas of dissent and conflict, and thus inform policy makers about the need and options for political intervention. In the early stages of technology development where debate is restricted to closed circles of experts, stimulating public debate can be necessary in order to avoid a narrow conceptual framework that over or under include problems. In case of vivid debates already going on in public, it can be a task of TA to stimulate the debate in the sense of expanding the scope of the actors involved in it or in the sense of contributing to it by expanding the scope of issues tackled, or social perspectives represented. This might be achieved by feeding in additional (unbiased) information or by facilitating transparent and balanced discussions by means of organising events with open access for any relevant actors and equal opportunity to put their arguments forward.

(h) Visions or scenario's introduced to actors

Scenario writing or other methods to explore future developments (e.g. "Delphi" procedures) are used frequently in strategic policy making and TA in order to cope with long-term developments and explore long-term effects of policies (e.g. the future of energy production and health care). These methods are applied to explore possible paths of policy making in a given complex situation and to establish consensus on measures to be taken.

These techniques can contribute to changing attitudes and opinions of actors by helping them to relate their perspectives and strategies to an expanded time horizon (long term effects of strategies and scenarios of social development they have to adopt their strategies to) or to explore alternative paths to achieve policy objectives by confronting them with new challenges or with perspectives and strategies from other branches of policy making. An impact of this kind is most likely to be achieved in a "context of discovery", i.e. in the first stages of discourse, when problem-definitions are not settled yet or a process of searching for operationalisation of common generic objectives or "Leitbilder" is going on (like e.g. in the case of "Sustainable Development"). Another case in point would be a situation where opinion making and decision making is paralysed and new perspectives are needed (see also "blockade running").

2.4.2.2
Mediation

The need for independent and well-balanced information refers not only to (inter-disciplinary) scientific analysis but also to the often conflicting and sometimes incommensurable opinions and interests held by societal groups. Providing for an ample picture of the actors' perspectives and involving every important group of actors in the TA process by open and transparent procedures can lead to a process of mutual learning about perspectives and interests. In turn this may induce new ways of communication and joint problem solving among actors.

(i) Self-reflection among actors

As a prerequisite to establishing common learning it is necessary that actors are able and willing to reflect on their own perspectives and interests. There is no way to force actors to do so, as long as they see opportunities to end up as a winner in a conflict (by means of power). However, TA procedures, by confronting the actors with criticism or forcing them to argumentatively substantiate their perspective (in the light of scientific results or societal values), may support readiness to modify one's own aims and expectations in order to make them more compatible with the needs of society or other groups. In local or regional planning processes there often appear manifest conflicts based on perceived violation of interests. These conflicts are often not (or not to a great extent) connected with fundamental value conflicts and debates about the appropriateness of normative frames (as is the case in debates of generic new technologies, e.g. biomedicine). In these situations participatory TA processes provide an arena where actors either are given the opportunity to negotiate aims and conflicting interests to find a compromise or to agree on compensation for violated interests. In the case of completely new technologies where there is uncertainty on effects as well as ambiguity with regard to ethical aspects, TA can induce self-reflection by establishing an arena of discourse – i.e. a procedure which provides for equal opportunities of articulation but restricts the modus of communication to arguments – thus "forcing" the participants to stick to standards of discourse ethics (Habermas 1991, von Prittwitz 1996).

(j) Blockade-running

Good chances for the actors to be willing to self-reflect usually arise when all the relevant actors see themselves caught in a deadlock. No actor may see a chance to get his or her position through without paying a high price in terms of economic costs or a massive loss in societal capital (image, trustworthiness, etc.). Or each actor defends a rigid political line so that no dialogue seems to be possible and no solution can emerge.

TA, as a non-partisan procedure, can contribute to blockade running by providing neutral ground for dialogue or by bringing in new ideas for problem-solving or a new problem definition which may help actors to reframe their position. It is necessary however that a commitment of all relevant actors to the TA or mediation procedure is achieved. Even though involved actors might agree on a common solution, this solution might be rejected by actors not involved in the mediation process. These actors might be other interest groups which, for various reasons, did not participate in the participatory process. The rejection can also come from organisations involved in the process, since not all their members went through the mediation and negotiation process.

(k) Bridge-building (trust-building)

One decisive prerequisite for blockade running and co-operation among actors is the establishment of trust. Mutual trust in technology debates is often a highly needed but at the same time always precarious resource of problem solving. Reasons for the lack of trust in technology controversies are manifold: conflicts about technologies often are based on mutually exclusive fundamental world views; technology risks are not equally distributed, so that particular groups see themselves as victims of technology policies whereas others benefit strongly. In those cases actors may see themselves as enemies with no overlapping interests that could provide ground for co-operation. TA can contribute to (re-) establishing trust by providing a platform where actors can meet and discuss at some distance from the public arena (where they usually stick to strategies of strongly promoting their interests). To succeed in being accepted as a mediator in these cases it is of course necessary that the TA-institute has established a non-partisan and competent image.

With growing importance of S&T for societal development the issue of trust (or lack of trust) in experts is moving to the centre of technology debate (Giddens 1991). Scientists often regard lay people's lack of understanding of complex scientific issues as being responsible for the loss of trust in expert-knowledge and growing criticism of experts. Social studies on science and technology however show that scientific knowledge and expertise undergo a process of "normalisation" i.e. no longer can lay claim to unquestioned authority. The more relevant expertise becomes for policy making, the more uncertainty and tentativeness of scientific knowledge with regard to societal demands of problem solving is revealed. It cannot therefore be expected to (re)establish trust in expert-knowledge (in the sense of unquestioned authority of science). New arrangements of dealing with uncertainty between science and society have to be found (Nowotny, Scott and Gibbons 2001). TA processes can contribute to this particularly with participatory procedures where scientists and non-scientists co-operate with regard to problem definition, interpretation and evaluation of data and knowledge in the light of societal demands and social perception of problems.

2.4.2.3
Re-structuring the policy debate

Changing the agenda of public and political debate – by introducing additional analytical knowledge and improving the willingness or ability of actors to reflect on vested interests and settled perspectives – may contribute to open up the policy making process by inducing new policy options. TA then can give way to reframing the debate on policy options – the urgency of the situation, the goals and the ways of political intervention.

(m) Comprehensiveness in policies increased and (l) policies evaluated through debate

It is often the case that technology policy debates are dominated by interests and perspectives of particular actors (industry, a scientific discipline). Dominant interests or dominance of a particular group of experts in public debate and policy making can be barriers to an open process of problem-definition and search for policy options. Administration routines and strict division of tasks and competencies among expert groups (or disciplines) may hinder an open debate on policy options. This restricts policy making to a particular problem-definition and accordingly to a particular perspective on problem-solving, disregarding other possible solutions (e.g. technical versus non-technical), side effects on other areas of society (e.g. economy – environment) or interests of other groups. This may cause sub-optimal strategies for problem solving and/or opposition against strategies by afflicted groups. By broadening the perspective, TA might help to avoid these problems. Comprehensiveness might be increased by taking rationales of several perspectives into consideration (social, economic, etc.), or by taking into consideration the viewpoints of a multitude of actors. The latter may lead to an evaluation of conflicting policy options by relevant actors that clarifies areas of consent and dissent among actors and give way to fine tuned policies with regard to different interests.

Notwithstanding the fact that providing "comprehensiveness" is one of the central tasks of TA in general, it may be mostly achieved by "problem-driven" TA projects, i.e. when scrutinising alternative ways of problem-solving is the focus of the TA-process. An example is the search for strategies to overcome traffic problems in a city/region. The discussion might be dominated by experts for technical solutions (build new roads, additional underground-line, etc.). A TA-process including other expertise for regional planning might show additional or alternative ways of problem-solving, avoiding negative effects (economic, ecologic) of the existing dominant strategies. It may become apparent that a long-term strategy for decentralisation of industrial and services or for a mix of working and housing areas could reduce traffic problems.

(n) Democratic legitimisation

Apart from providing for comprehensiveness with regard to contents (description of the problem, un-intended consequences analysed, perspectives considered) a TA process can contribute to policy debates on a meta level of "political culture" when the openness and fairness of the process succeeds to establish the notion among actors that the decision making process met the rules (or ideal) of open democratic deliberation. This role can also be labelled "Legitimisation by process" (Luhmann 1969). Often controversies on technological developments and their implementations are caused by fundamentally opposing interests and ethical perspectives of actors. Usually then there is no "win-win" solution or consensus possible. A TA process can at least ensure that the perspectives of all relevant interest groups have been acknowledged and appropriately scrutinised. So, even if some actors might not agree with the results of the TA-process or the respective conclusions and decision of policy makers, they might still acknowledge that the process has been "fair" and "democratic".

In case of very contradictory positions in political debates, however, the willingness to take part in the TA-process as well as the willingness to accept the TA process as a means for democratic decision making may be weak. It is known from participatory TA procedures that stakeholders may regard commitment to the TA-process as being restrictive to their strategic position in the political debate. Commitment to a formal process of "discourse" can (from a stakeholders' perspective) be seen as "dangerous" since the outcome cannot be predicted (van den Daele 1995). So the process may be blamed with hindsight for not having been democratic.

A general problem in this context is that actors (in most cases those with weak power in formal decision making processes) often connect their participation in a TA-process with the expectation to directly influence decision making. In cases where they cannot see any influence on established decision making processes, the TA-process is criticised as being "pseudo-democratic".

The commitment to and acceptance of the process as a contribution to democratic decision making is more likely in cases where a willingness to "joint problem solving" already is established among stakeholder groups (e.g. because the issue is new and positions are not fixed yet, or because all stakeholders feel the need to overcome a blockade in decision making).

2.4.3
Initialising action

Impact in the dimension of "initialising action" means that a TA process influences the outcome of the policy making process. Regarding the scientific aspects of the issue at stake a TA-process may lead to *new R&D policies*, i.e. initiatives to further scrutinise aspects of the problem (e.g. a research programme to explore the risks of deliberate release of GMOs). With regard to the societal aspects (actors, conflicts) policy makers may conclude from a TA-process to initialize *new ways of decision making* (e.g. to set up a programm of public discourse or include social groups in the decision-making process). Apart from such initiatives which can be seen as new forms of dealing with a problem it might well be that TA leads to a definite *political decision* (in the sense of closure of debate): e.g. to implement a technology that was scrutinised with regard to its pros and cons, or to set up legal rules for implementation.

In a way, the impact dimension of "initialising action" represents the end of the policy-making process, or the (provisional) end of a policy making cycle (which often at the same time marks the start of a new one). Taken properly it is somewhat inaccurate to speak of a "role" for TA to play in this dimension. Whereas TA "plays a role" in the realm of the other two impact dimensions (i.e. having the mission as well as tailored methods for raising knowledge and intervening in the process of opinion forming) there is no mandate for TA to directly take part in decision making in the sense of doing politics. Quite the contrary: policy makers would regard it as an illegitimate intervention, if TA tried – as it were – to prescribe what kind of decision should be regarded as appropriate given the achieved state of knowledge and debate so far. Nevertheless TA is expected to contribute to policy making and thus to make a difference in terms of having an impact on decisions taken with regard to the issue at stake. However, since TA takes a role in raising knowledge and organising debates, it is a visible player in this process, whereas TAs influence on decisions taken is much more unlikely to be observable and is much more dependent on a lot of intervening factors.

2.4.3.1
Reframing the debate

An implicit expectation held by policy makers with regard to TA is that in-depth analysis of the problem at stake will reduce complexity of the related decision making problem by sorting out what is "right" and "wrong" and thus showing which policy options are reasonable to go for. In some cases this may be achievable. More often however analysis of the debate amongst experts on topics such as risk assessment or the costs and benefits of a particular technological option, reveals the complexity of the problem and the inherent uncertainty of scientific knowledge and thus expands the frame of problem perception. One reaction to this may be to start an initiative to further scrutinise the problem from a new perspective and hopefully establish more secure ground for decision making. TA then contributes to initiatives by policy makers or other actors to continue the assessment of the problem on a new level of knowledge about the "factual" or "material" aspects of it.

(o) New action plan or initiative to further scrutinize the problem at stake

R&D policy making has to deal with high degrees of uncertainties with regard to future outcomes, while at the same time usually stakes are high (investments, opportunity costs) and decisions are urgent (Funtowicz/Ravetz 1992). New technologies or research areas are often promoted by powerful scientific communities that might over-estimate chances and under-estimate risks. TA has the mission to broaden the discussion, include views of other expert communities, and give a comprehensive view on possible effects of implementation of an innovation.

When a TA process has made existing lack of knowledge or dissent among experts visible, or revealed new problems that had not been explored yet, this might lead to initiatives to further explore the problem at stake. Initiatives can be taken by expert communities or policy makers. This may be a new research project or a new (additional) TA-project on the issue, a new R&D funding programme, or the setting up of an expert committee.

Opportunities for TA to induce an initiative to continue or intensify "learning on the issue" are usually high when the different interests are not settled yet – in particular when technology development is at an early pre-market stage. Another window of opportunity might be given when decision making processes are blocked by competing major interests (lack of public acceptance against interests of industry to take the pole position in international competition). Then a provisional solution might be to "wait" for further scientific clarification and get the chance to postpone decisions. On the other hand, when investments in innovation have already been taken, it is not likely that policy makers (or other actors) see the need or chance to further scrutinise central aspects of the issue and prolong the debate.

(p) New orientation in policies (aims, objectives)

TA can contribute to re-orientation of policy making in terms of initiatives to explore new objectives. Opportunities to do so might arise when relevant actors are looking for new solutions or a common definition of a problem. This is the case when a technology is quite new and its effects are not yet foreseeable. TA, by supplying comprehensive analysis, might contribute here to a shift in the framing of the policy process (e.g. from addressing a technology from the perspective of industrial development to a perspective of avoidance of risk and application of the precautionary principle – and vice versa).

Another case in point might be the ongoing discussions on general orientation in a policy branch; for instance how to apply the model of "sustainable development". "Re-orientation" towards new long-term objectives and adoption of new aims in strategies (for instance in a ministry) may be supported by TA via exploration of the instruments needed to implement the new objectives as well as compatibility with existing aims and strategies.

2.4.3.2
New decision making processes

Apart from giving rise to initiatives to continue debate on a new level of knowledge about scientific ("factual") aspects of an issue, it is often the case that the exploration of structures and problems in the debate lead to initiatives to restart (or continue) debates on a new level of inclusion of relevant actors or to apply new procedures of negotiation or bargaining among relevant actors.

(q) New ways of governance introduced

TA activities such as consensus conferences or other procedures of public participation may induce initiatives to involve actors in a more direct way in decision-making than through classical democratic representation. These initiatives may be embedded in specific programmes of new governance (top down) or are experiments that grow from the need of specific interest groups to bridge the gap left by the perceived deficiency of established democratic channels. Important when considering this role/impact is that the need for the introduction of new ways of governance is often not only driven by problems concerning new technologies but mainly by the perceived need for involving the general public or specific societal groups in a different and more effective way in processes of decision making – as it is dealt with in the debate on "Governance of Science" and the need for new ways to integrate science and society (Fuller 1999).

(r) Initiatives to broaden, intensify, stimulate public debate or dialogue among actors

The analysis of differing interests and perspectives with regard to the issue at stake (given by what has been called here "social mapping") and moreover attempts to mediate between conflicting actors may convince them that there is a need to include a broader range of afflicted interests into the debate, or to expand it beyond circles of experts, or to raise public awareness on the issue and get to know more about ethical perspectives of the general public. Especially where there is a high degree of uncertainty about legitimate or socially acceptable decisions, this may motivate policy makers (as well as expert communities) to start an initiative or set up a program to intensify or broaden the debate. New initiatives to stimulate public debate may embrace conventional means like PR-campaigns or funding of consensus conferences (or similar types of activity). Another example would be the establishment of a new advisory committee that includes representatives from interest groups that so far were not involved in the process of policy making.

2.4.3.3
Decision taken

To initialise new policies in terms of taking a decision on the issue at stake of course can be regarded as the ultimate impact of any process of policy consulting. However, it is well known that the road from knowledge to political action (which is the road from "policy analysis" to "new policies" in the typology) is by no means a straightforward one. Any scientific policy advice as well as any result from participatory procedures of policy consulting (DeLeon 1990) undergoes a lot of filters (made up by interests and perspectives of actors influenced by other resources). Consequently, the specific impacts or roles of TA mentioned below are highly mediated ones and not directly related to TA's mission. Most of the roles addressed in the typology can be – mutatis mutandis – understood as "goals" which a TA institute may follow when setting up a TA process. Those roles mentioned under the dimension of initialising action and in particular those that are categorised under "new policies" can not be regarded as "goals" of a TA project.But they should be regarded as impact in the strict meaning of influence on the outside world that TA might have but are not in the reach of TA itself (its methods and activities, the design of a TA project, the quality of outcome, etc.)

(s) Filter of policy alternatives

To a certain extent discussion and evaluation of policy alternatives arepart of most TA processes. Policy alternatives are scrutinised with regard to their practical viability and their economic, social and environmental effects. Both expert and participatory TA are well placed to perform this role since it includes both technical details as well as value judgements. Evaluation of policy alternatives is intended to support and sometimes leads to conclusions about which policies should be implemented and which not. This may not necessarily result in legislative activities but may result in the shaping of governmental R&D programmes and allocation of funds with regard to different objectives.

When evaluating policy alternatives – which are often based on diverging interests and values being held by stakeholders – it is crucial to avoid politicisation of the TA process. TA needs to be particularly transparent throughout the process in terms of choice of experts and topics, as well as fully inclusive of major political representations.

(t) New innovation-process implemented

Technology development is largely market-driven (supply push / demand pull). The market is however dominated by economic criteria: not demand as such, but well funded demand decides on the social diffusion of technology. Whether an innovation is socially (with regard to values and needs) and environmentally sound or not can in most cases only be stated in hindsight. R&D policy can be understood as an answer to "market failure". There is a need to steer technology developments according to societal needs that the market might not care for. This includes the attempt of "social shaping" of upcoming new – supply push driven – technologies as well as to trigger development of technologies that are socially needed or useful (demand pull) but not supplied for by industry. In both cases R&D policy needs knowledge about the spectrum of possible technological solutions as well as social needs and demands.

TA can contribute to the shaping of technology according to social needs and thus facilitate the introduction of technologies by inducing related R&D development programmes. These may be technologies that are alternatives to existing technologies, but which live up to better standards (i.e. technologies that deliver better working conditions; sustainable solutions; low risk technology shifts etc.). This can be regarded as one of the fundamental ideas behind the consept of TA and has made up the focus of the Dutch concept of "Constructive TA" (Rip, Misa and Schot 1995). Typical examples are programmes – set up in most European countries – to promote the development of "environmental technology" in the meaning of technology that inherently lives up to objectives of environmental protection (i.e. reduces the consumption of natural resources as well as the environmental load in terms of waste and environmental pollution).

The opportunities to have an impact on innovation processes are highly dependent on the industrial structure: Are there important branches that support new technologies? Or on the other hand: Are there important branches that see the new technology under consideration as a menace (new technology might substitute their products)? It is important whether win-win-solutions are possible (double dividend). The public debate about risks as well as about the need for better technological solutions is important. Are there influential NGOs which support the new direction of technology development? Is the wider public (the media) attentive with regard to the problem?

(u) New legislation

To initialize or influence legislative action i.e. legal regulation with regard to a technology or an issue at stake probably represents the ultimate impact of any process of policy advice. When advice itself becomes policy, it has succeeded in providing the most comprehensive and pragmatic solution to the issue in question.

Nevertheless, TA's direct role is not, and could never be, to create policies. To develop and explore options for policy making includes input to the preparation of legislation. It might well be that policy options developed by a TA-process give particular hints (advice) to aspects that have to be considered in preparation of legal rules. Most TA studies e.g. carried out during the 1990s in the field of genetic engineering or biomedicine (such as genetic testing) were carried out in a context of discussions on legislative stop and go decisions for the application of technologies. Those studies often explicitly dealt with the pro and cons, the opportunities and restrictions and possible features of legal regulation, but did not intend to prescribe legislation by providing a (draft) bill.

The exploration of the need and opportunities for legal regulation carried out by TA can only take up, scrutinize and (re)structure arguments and demands expressed in public and policy debates. And policy making in general – but legislative activities in particular – is structured and influenced by a lot of players in the field and by difficult and sensitive processes of negotiation among actors and policy networks in the forefront of rule making. TA here is nothing more than one – albeit independent – voice and source of information amongst others. As with the impacts of TA on policy making in general it becomes quite clear that the best role for TA to play is to make a full picture of the pros and cons etc. available to decision makers and other actors. Therefore it wouldl hardly be expected to be possible to track a clear line from a TA-process to the specific features of legal rules that have been decided upon.

2.5
Influencing factors

As it has been described in the introduction to this chapter, measuring the overall impact of TA is a rather complicated process as it involves separating the TA process from a large array of other parallel processes that affect the final decision making. Whether TA has been successful in its mission and has achieved its goals, depends both on internal institutional decisions to do with correct situational identification and choice of methodology (see method paper), and on external factors that TA might have little control about but should nevertheless take into consideration.

Influencing factors on the impact of TA, other than the chosen methodology, denote limits in the conception and execution of the TA process and at the same time present formidable challenges in the final reception of the TA study. The TAMI group has identified three main categories of influencing factors: Institutional Setting, Technology Policymaking Culture, and Structure and State of the Innovation Process. In terms of the typology of impacts as it is described in the previous sections these factors relate best to the "issue dimension", where, "Institutional Setting" refers mainly to the dimension of "technological/scientific aspects", "Technology Policymaking Culture" to "societal aspects", and "Structure and State of the Innovation Process" refers mainly to "policy aspects". Following is a brief description of the influencing factors.

2.5.1
Institutional setting

The particular organisational structure of the TA institute naturally poses certain limitations on the type of work it can undertake and the manner in which this can be done. The overall mission of the organisation and the main target groups for its work constitute the first such limitation; further limitations can be identified as the official relationship between the institute and its "customers", the institutional image and overall state of competition.

Parliamentary vs. non-parliamentary setting

The TA institute could either be attached to the national legislature (e.g. TAB, POST) and thus be a "parliamentary office" or have a more "independent" status of a research– or academic institution (e.g EA, ITAS). This difference does not necessarily affect the independence and value of its work, but nevertheless denotes certain limitations in the work process due to the main target audience. Such limitations will doubtless have an effect in the perception and eventually the impact of the institute's work.

There is great variety in the setting, organisation and overall mission of parliamentary offices in Europe. Reflecting perhaps the general national decision making cultures, the mission of such institutes ranges from strictly adhering to the needs of the parliament (e.g. POST in the UK) to having a more general role in promoting public debates and acting as a bridge-builder in socially sensitive issues (e.g. DBT in Denmark). Despite this diversity which itself poses direct limitations in the work of the institute, there is a common denominator in the fact that the main audience of the institute is always the policymaking community and the main workshould therefore focus exclusively on their needs. The limits this institutional arrangement poses include adherence to strict political neutrality, fast-track analysis (due to policymaking time constraints and policymakers' low attention span), and sometimes low institutional visibility (since parliamentary offices are mere tools in the policymaking process)[9].

Non-parliamentary institutional settings are relatively rare in Europe (with the exception of Germany) due mainly to budgetary constraints. The few existing institutes enjoy a relatively more independent institutional arrangement since their mission and target audiences are usually described in general terms. This provides considerable flexibility in terms of working manner, study timing and results presentation since the work is independent from direct policymaking processes. In addition, there is usually self-determination in promoting the institute and raise public awareness of its work. Conversely, the institutional setting does not guarantee access to the policymaking community (which the typology identifies as a main impact dimension under the title "initialising actions") and such institutes are also very vulnerable to general trends on budgetary cuts and public resource re-structuring[10].

Reactive vs proactive setting

This refers to organisational decision making structures relative to the topics chosen for studying. A reactive organisational structure depends on external factors when choosing studies, the time allocation for the conclusions and even the preferred process. It refers to a situation where the TA institution is requested, or proposed as being the appropriate agency, to conduct an investigation (e.g. in the parliamentary context, by a parliamentarian in a debate, or by a parliamentary committee in the course of an inquiry). Whether or not the institute is obliged to follow or has some freedom of choice in the requests given does not alter the fact that its organizational structure is basically reactive. An example of purely reactive institutional settings is STOA at the European Parliament which consists of a secretariat

---

[9] For a detailed account of European Parliamentary TA settings, see "Shaping the impact: the Institutional Context of Parliamentary Technology Assessment" at the Supplement Papers section of this book.

[10] The recent closure of the TA Academy at Baden-Württemberg, probably the biggest European TA institute, for budgetary reasons is a sad reminder of the uncertain future for non-parliamentary TA institutes.

that receives requests by members of the parliament and outsources the studies. On the contrary, a proactive organisational structure refers to a situation where the TA institution decides, through its own internal programme-setting procedures, to conduct a study. This is a direct structure that maximizes self-determination and resource planning although this freedom is restricted by the fact that there is always a main target audience that requires relevance. An example of a purely proactive type of TA institute is the Europaeische Akademie GmbH in Germany that plans its own work programme. This programme needs to be accepted by its funding organizations.

Usually there is no clear-cut distinction of TA institutes along the "reactive/ proactive" category. Both parliamentary and non-parliamentary institutes might have either a proactive or a reactive structure, or indeed a combination of both. In either case there are limitations involved: in the reactive structure there is little flexibility in the choice of subjects and methods while in the proactive case the work can be of little policy relevance. One might argue that strict adherence to either type tends to influence the work outcome and eventual impact, and should therefore be undesirable.

Institutional image

As with every other service, TA is influenced by competition forces (see also section on "organised interests" beneath). The number and diversity of TA services (e.g. risk assessment, technology foresight, bridge building, etc.) as well as the limited number of topics at any given time with direct society or policy interest, raises competition amongst TA actors and institutions. The image and public standing of the institutes will inevitably play a major role in target audience preferences and subsequent effect of the TA study.

The image of TA institutes is a rather elusive topic as there is virtually no comparative research in Europe as to the level of European TA institutes' public standing or the elements of their public image. Similarly there is no "ready-made" recipe for success in this area. The institutional quality control processes differ immensely across Europe and offer no guarantee of raising public standing. As it is nevertheless a vital aspect for the future development (some might say survival) of TA to have a coherent and fair approach to improving its own image, some urgent research is requested in this area.

2.5.2
Technology policymaking culture

Organised interests

The power of organised interests in influencing policymaking can hardly be underestimated. The number of different interest groups represented in Brussels alone more than tripled between the mid-80's and mid-90's. The interests represented range from individual companies and European interest associations to NGOs and trade unions, while the number of individuals involved directly in lobbying activities is estimated at around 10,000. Various studies have corroborated the strong influencing power of organised interest groups. They show that the great majority of policymakers at the European level receive information mainly from interest groups and lobbyists. Moreover, interest groups more frequently provide direct voting instructions to members of the European Parliament than either the party leadership or the national governments.

Within this political reality TA is faced with tremendous competition as a policy advisory service. This competition could furthermore be considered "unfair" given the amount of resources that lobby groups usually are able to use in promoting their message. Despite the fact that TA can usually claim independence from interests and consequently greater validity in reporting, the limited time that policymakers can afford in information intake often means lower priority is given to bulky TA studies in favour of more focused and politically aware information given by various interest groups[11].

Public awareness/level of social debate

The level of the current debate in society on technology issues is another uncontrollable influencing factor on the overall impact of TA. Media attention on particular scientific issues or technological discoveries can affect dramatically not only the need for immediate policy advice but also the themes and overall orientation that this advice should take. At the same time, high public awareness makes it more likely that policy makers accept the broad range of information and policy options delivered by TA. This is the opposite of low levels of public awareness where policy making is more receptive to organised interests promoting the technology at stake.

Unfortunately, TA is poorly equipped to deal with sudden turns in public debates as it often enrols in a study process that can hardly change during implementation. At the same time TA rarely functions as a communication node and therefore does not have the means to follow or take part in delicate social debates. The implicit requirement for more flexibility during the TA process is very significant for the relevance, and therefore effect, of the TA advice. This will require re-evaluation of current TA methods and appropriate alterations to make the TA process sensitive and responsive to sudden changes in the subject matter[12].

2.5.3
Structure and state of the innovation process

State vs. market driven innovation

The way that the overall innovation system functions presents another factor affecting the impact of TA. No two national innovation systems are identical in Europe but there is certain similarity in indices such as R&D expenditure, share of private sector in national R&D, priority technology sectors, etc. These rough similarities nevertheless break down when considering specific technology subjects where a variety of innovation structures and trajectories appear in the picture. A general distinction of the innovation processes in Europe can use the categories state vs market driven.

---

[11] The extent of the problematic of lobbying activities to the work of TA is analysed in detail in "Organised Interests in S&T Policy; The influence of lobby activities" in the Supplement Papers section of this book.

[12] For a detailed account see chapter on "The Practice of TA; Science, Interaction, and Communication".

In state driven innovation systems the prime driving force behind the adoption of new technologies are public fora. Whether due in large part to the public interest in a particular technology or due to significant amounts of public funds spent on them, the government usually has a big say in the introduction of new technologies. In such cases, TA can have a more direct influence in policy making and the shaping of technology since it is an accredited public service, free of interests and agendas.

In market driven innovation systems, most technology development is market driven and the market decides whether a technology is acceptable or not. Constitutionally there is no legitimisation for the state to intervene in technology development on the basis of whether there is a societal "need" for this technology or not. Legal regulation can only come in if there is a high probability of harm to health or the environment (freedom of research, private rights of enterprises).

State of Innovation

The "Colingridge dilemma" accurately depicts this influencing factor that TA is often faced with. It relates to the notion of timing in the innovation process and how that affects any review of new technologies. It states that the earlier TA enters the innovation trajectory, the more possibilities there are to shape the future of the technology at stake but at the same time the more partial and vague the available information is. On the other hand, the later TA enters the trajectory, the more complete and comprehensive the knowledge over the technology is, but at the same time there is less chance to influence the innovation strategy.

There is no rule of how to solve this problem. Timing is of utmost importance for the overall impact of the TA process, but this might also be a factor where there is very little institutional influence. A constant flow of expert information is essential for the TA practitioner to decide whether the time is ripe for a study, provided that the institutional setting allows for internal decisions and the TA process is flexible enough to take into account new emerging themes and topics (see above).

2.6
Conclusions

The typology and the set of roles developed by the TAMI project show that, regarding the impact of TA, there is more to be considered than the influence of TA on political decision-making (i.e. the lower right box of the typology). TA – whether "classical" or "participatory" – contributes to social debates and policy making in many ways: by supplying unbiased knowledge, by supporting communication processes, by offering new perspectives on the issue at stake, by opening up new opportunities to restart debate in deadlock situations etc. The matrix of roles provided by the typology can support reflexivity in the design of a TA project (see also "The Practice of TA; Science, Interaction, and Communication" chapter) since it provides a structure for clarifying the goals of a project. In this respect the typology may also function as a means to clarify what kind of support for policy making in a particular case is desirable – e.g. relating to lack of knowledge on policy options, uncertainty about scientific data, lack of trust among stakeholders, deadlock of policy debate, etc. This is not to say that a TA institute is always free to choose which goal or role to go for. In most cases there are many restrictions set by the mission, the institutional setting or the organisational competencies of a TA institution, as described above. It is also obvious that the social and political environment restrict the choice of roles and the opportunities to influence the debate. In any case, however, reflection on the roles TA can play will support the situation appraisal and lead to awareness of what realistically can be achieved and what scientific, procedural and communicative methods may be applied and arranged in the project.

Given the complexity of the issues TA has to deal with, as well as the complexity of the political environment TA is embedded in, it cannot be expected that recommendations can be given about which particular methods should be applied in order to achieve certain impacts. In most cases a mix of methods might be most appropriate. As well it might not be possible to strictly aim at any one specific role.

However some more general insights into the relationship between impact/role and method have been triggered by making use of the TA typology: Given the two TA paradigms – the classical scientific policy consulting model and the participatory TA model – it can be said that the methods these models typically draw upon are strongly (but not exclusively) related to one dimension of the impact typology. The methods applied by a policy consulting type of TA are mainly scientific; it is the core mission of those institutes working exclusively on behalf of a parliament to support decision making by providing knowledge on scientific, social and policy aspects of the issue at stake and this is done mainly by drawing on the state of the art research. This also implies that the TA practitioner does not directly interfere in ongoing debates but rather observes these debates from a distanced, impartial point of view in order to give a comprehensive and unbiased analysis of the problem. Quality criteria such as argumentative quality and transparency of the analysis can be regarded as prerequisite for the acceptance of TA results by the client or other potential users. It is by the (scientific) quality of the product (albeit reflecting the quality of the process) that TA intends to make the user aware of problems and arguments not known so far.

Impact in the second dimension, i.e. in the sense of mediating conflicts, can be furthered by including stakeholders in the debate and providing new ground for mutual learning via application of participatory TA methods. It is however not to say that an unbiased, scientific analysis of the issue and of the stakes and interests implied in debate can never induce self-reflection among actors and thus support conflict resolution. However in participatory TA the scope of methods applied is expanded in the direction of direct intervention in the debate – i.e. not only providing knowledge but also organising and stimulating the debate. Although references to scientific knowledge are an integral part of this process, it is the interactive quality of the process (fairness, openness to relevant actors, etc.) that gains importance here. The TA organisation is still acting on the basis of the principle of impartiality, but the organisation and the TA process itself are more part of the debate than is the case with classical TA – or even more: the TA process is delegated to the social actors while the TA organisation only functions as a facilitator.

It can be said that classical TA is – with regard to its methods – related to the first impact dimension of the typology and participatory TA to the second dimension, and this should obviously be taken into consideration when evaluating the impact of

TA. This however should not be taken too narrowly – in the sense of each of them being restricted to their respective type of impact.

Regarding the mission and the methods of TA it is evident that "raising awareness" and "forming attitudes and opinions" are in the reach of both classical as well as participatory TA. Since TA has been regarded from the beginning as an answer to the growing need for comprehensive scientific information for policy making as well as to the ongoing debates on the impact of science on society, TA has always been regarded as implying comprehensive scientific analysis as well as taking account of different social perspectives. The increasing demand for and application of participatory (process related) methods in recent years should be regarded as a reaction of TA to a change in its social and political environment, i.e. the growing demands of civil society to be involved in policy making on scientific and technological issues and the increased need of policy makers to cope with dissent and conflicts on emerging technologies.

If TA expands the scope of methods towards intervening directly in the process of opinion forming, this means intervention by providing new processes of communication and debate. This does not imply to induce a particular opinion, perspective or a particular political position into the process and thus to directly intervene in the third dimension of the typology: the decision making process itself. In a way this is not surprising since it is part of the mission and self-conception of TA to obey a strict separation between policy making on the one hand and policy consulting on the other. However self evident this may be, it is important to stress that whereas TA has methods at hand to raise knowledge about scientific and social perspectives on issues as well as to organise, stimulate and moderate communication among actors, having a direct impact on decision making (often seen as being the ultimate goal of TA) is beyond the reach of TA's "method toolbox". It might even be regarded as being somewhat misleading to describe "roles" to play for TA in this context and instead we should be speaking of "contingent" outside effects of a TA process that are dependent solely on the state of the political process itself. Discussions among the TAMI project group on the importance of "communication" as a means to expand the opportunities for TA to produce impact (see "The Practice of TA; Science, Interaction, and Communication" chapter) raised the question of whether there is an opportunity and a legitimate chance for TA to take a step further in intervening in debate by adopting a more active role as a "player" in debates on science and technology. Whereas some pointed to the danger of a possible loss to the view of TA as an impartial observer and facilitator of debate and decision making, it was regarded by others as necessary for TA to be visible in the media as an actor in its own right, for example by taking the role of an "agent provocateur", challenging opinions and perspectives of actors in order to intensify the interaction between civil society, policy makers and scientists. In this case TA would give up its traditional role of a "knowledge broker" and its affiliation to the branch of scientific policy consulting and move towards the role of an advocate or activist to provide civil society with a say in technology debates. The scope of actual concepts of TA in Europe seems to be marked by this position (e.g. currently discussed in the Netherlands) on the one hand and the understanding of TA as being exclusively dedicated to proliferation of best scientific knowledge available to policy makers on the other hand. However, the practice of TA in most cases is likely to fall in between these categories and there is constant discussion in the TA community about which way to expand or adjust the set of methods (with the aim of improving the impact on policy making and social debate) and what this will mean with regard to the mission and competencies of TA institutes. This ongoing discussion on self-understanding, function and impact of TA is ultimately solidified by the typology of roles and impacts developed in the TAMI project.

References

Albaek E (1995) Between Knowledge and Power. Utilisation of Social Science in Public Policy Making. Policy Sciences 28:79–100

Beck U (1985) Risikogesellschaft. Suhrkamp, Frankfurt a.M.

Berg MR, Michael DN, Brudney JL (1978) Factors Affecting Utilisation of Technology Assessment in Policy Making. Centre for Research on Utilisation of Scientific Knowledge. University of Michigan

Bütschi D, Nentwich M (2002) "The Role of Participatory Technology Assessment in the Policymaking Process". In: Joss S, Bellucci S (eds) Participatory Technology Assessment – European Perspectives. Westminster University Press, London 253–256

Caplan N (1979) The two Communities Theory and Knowledge Utilisation. American Behavioral Sciences 459–470

Cohen MD, March JG, Olsen JP (1972) A Garbage Can Model of Organizational Choice. Administrative Science Quarterly 17:1–25

DeLeon P (1990) Paricipatory Policy Analysis: Prescriptions and Precautions. Asian Journal of Public Administration 12:29–54

Fuller S (1999) The Governance of Science. Buckingham, Open University Press

Funtowicz SO, Ravetz JR (1992) "Three Types of Risk Assessment and the Emergence of Post Normal Science". In: Krimsky S, Golding D (eds) Social Theories of Risk. Westport 252–274

Giddens A (1991) The Consequences of Modernity. Polity Press, Cambridge

Habermas J (1991) Erläuterungen zur Diskursethik. Suhrkamp, Frankfurt

Hennen L (1999) Participatory Technology Assessment – A Response to Technical Modernity?. Science and Public Policy 26:303–312

Hennen L (2002) "Impacts of Participatory Technology Assessment on its Societal Environment". In: Joss S, Bellucci s (eds) Participatory Technology Assessment – European Perspectives. Westminster University Press, London 257–275

Héretier A (1993) Policy Analyse. Elemente der Kritik und Perspektiven der Neuorientierung. Politische Vierteljahresschrift, Sonderheft 24:9–38

Hetman F (1973) Society and Assessment of Technology. Paris: OECD

Jasanoff S (1990) The Fifth Branch: Science Advisers as Policymakers. MA: Harvard University Press, Cambridge

Joss S, Bellucci S (eds) (2002) Participatory Technology Assessment – European perspectives. Westminster University Press, London

Kingdon J (1984) Agendas, Alternatives and Public Policies. Little, Brown, Boston

Liakopoulos M (2001) The Politics of Technology Assessment. In: Decker M (ed) Interdisciplinarity in Technology Assessment; implementation and its chances and limits. Springer, Berlin

Luhmann N (1969) Legitimation durch Verfahren. Suhrkamp, Frankfurt a.M.

National Academy of Sciences (1969) Processes of Assessment and Choice. National Academy Press, Washington DC

Nowotny H, Scott P, Gibbons M (2001) Rethinking Science – Knowledge and the Public in an Age of Uncertainty. Polity Press, Cambridge

OECD (1983) Assessing the Impacts of Technology on Society. OECD, Paris

Paschen H, Petermann Th (1991) "Technikfolgen-Abschätzung. Ein strategisches Rahmenkonzept für die Analyse und Bewertung von Techniken". In: Petermann Th (ed) Technikfolgenabschätzung als Politikberatung. Campus, Frankfurt and New York 19–42

Paschen H, Bechmann G, Coenen R, Franz P, Petermann Th, Schevitz J, Wingert B (1991) "Zur Umsetzungsproblematik bei der Technikfolgenabschätzung". In: Petermann Th, Technikfolgenabschätzung als Politikberatung, Campus, FrankfurtNew York 151–184

Rip A (1986) Controversies as Informal Technology Assessment. Knowledge, Creation, Diffusion, Utilization 8:349–371

Rip A, Misa Th J, Schot J (eds) (1995) Managing Technology in Society – The Approach of Constructive Technology Assessment. Pinter, London New York

Scharpf FW (1993) Positive und negative Koordination in Verhandlungssystemen. Politische Vierteljahresschrift, Sonderheft 24:57–83

Sabatier P (1988) An Advocacy Coalition Framework of Policy Change and the Role of Policy Oriented Learning Therein. Policy Sciences 21:129–168

Torgersen D (1986) Between Knowledge and Politics: Three Faces of Policy Analysis. Policy Sciences 19:33–59

Van den Daele W (1995) Technology Assessment as a Political Experiment. In: von Schomberg R (ed) Contested Technology, International Centre for Human and Public Affaires, Tilburg/ Buenos Aires 63–90

Van Warden F (1992) Dimensions and Types of Policy Networks, European Journal of Policy Research 21:29–52

Von Prittwitz V (ed) (1996) Verhandeln und Argumentieren. Leske und Budrich, Opladen

Vig N, Paschen H (eds) (2000) Parliaments and Technology – The Development of Technology Assessment in Europe. State University of New York Press, Albany

3 Technology Assessment in Europe: Conclusions & Wider Perspectives

Lars Klüver, Sergio Bellucci, Robby Berloznik, Danielle Bütschi, Rainer Carius, David Cope, Laura Cruz-Castro, Michael Decker, Søren Gram, Armin Grunwald, Leonhard Hennen, Theodoros Karapiperis[13], Miltos Ladikas, Petr Machleidt, Luis Sanz-Menéndez, Wim Peeters, Jan Staman, Susanne Stephan, Tomasz Szapiro, Stef Steyaert, Rinie van Est

3.1 Introduction

This chapter presents the lessons learnt, with an emphasis on the implication and scope of the TAMI project results. It draws on the work of the two TAMI groups presented in the previous chapters "The Practice of Technology Assessment; Science, Interaction, and Communication" and "Towards a Framework for Assessing the Impact of Technology Assessment" and presents overall conclusions and recommendations.

The project TAMI had the aim of creating a structured dialogue within the Technology Assessment (TA) community as well as between TA experts and policy makers in order to improve our understanding of the effect of TA in science and technology policy-making. In spite of its reputation as one of the most self-reflecting research communities, TA traditionally has only spent few resources on evaluation of the impacts of its work. There may be many reasons for this, but it surely is of importance that the effects of such indirect advisory functions are very hard to document. TAMI therefore aimed at answering two main questions:

- What kinds of impacts of TA can be identified and how can these be categorised?
- How can TA optimise their activities with regard to supporting science and technology decision-making processes in our societies? And how can the choice and use of methods contribute to such optimisation?

The partners of the TAMI project include researchers from Belgium, Czech Republic, Denmark, Germany, Netherlands, Poland, Spain, Switzerland, and United Kingdom. These participants include the known variety of institutional settings and methodology. Further, the participants have included members of the EPTA Network[14] and the EUROPTA Project[15], in order to gain advantage from the discussion of impacts and methods that have taken place in these professional communities.

---

[13] "The views expressed are those of the author and do not necessarily reflect those of the European Parliament".
[14] EPTA is the European network for Parliamentary Technology Assessment offices. See www.eptanetwork.org
[15] The EUROPTA project examined participatory technology assessment methods and scrutinized 15 case studies with regard to implementation and choice of method, project management, political roles and impact. The report and case studies can be found at www.tekno.dk/europta. Footnote 3 gives reference to the EUROPTA book.

The starting point of TAMI was the international development of TA, which has left TA practitioners with a broad experience, a growing theoretic and methodological understanding and an increasing field of application. But also, TA is a relatively young discipline, still developing its aims, tools and self-understanding. TAMI may be seen as a project that has focused on the most important fields for clarification on the functions of TA: intervening into the right problems, picking the right tool, at the right moment, and with the right effect.

3.2
Three central terms

Analysing the relation between method and impact is in essence an exercise looking at the functional relationship between TA and its surroundings. This has made it necessary for TAMI to revisit the definitions of *technology assessment, method* and *impact* in order to ensure a common functional understanding of these three basic terms.

Technology assessment has been subject to many definitions, focusing on certain aspects of TA, such as for example the aims, the methods, or the addressees. Though – or maybe because – such definitions serve specific aims for different purposes, none of them seem to cover the functional relationship between the full range of TA approaches, and the full range of outcomes. The following definition was therefore developed in order to be able to embrace the processes and aims involved in TA (see the chapter "The Practice of Technology Assessment; Science, Interaction, and Communication"):

> Technology Assessment is a scientific, interactive and communicative process with the aim to contribute to the public and political opinion forming on societal aspects of science and technology.

A *TA method* is not a unit, which is easy to pinpoint. Methods can be described as the more or less well-documented procedures in the TA toolbox. But such an approach does not embrace the dynamic methodological process inside a TA project. Neither does it include the relationship between these processes, the goals of the TA activity, and the societal situation surrounding the project. Consequently,

> TAMI has interpreted methods as a plurality of scientific, interactive and communicational procedures, involved in a TA project design.

Such procedures may formally be documented, or may be present in terms of the competence of the project organisation.

Impact of TA traditionally refers to the expectation, that TA should increase the quality of decision-making, which is often connected to a demand for influence on decision-making, and the idea that real impact occurs when political decisions have been taken. However, this understanding of impact suffers from the problem that it simplifies and to some extent ignores the nature and reality of decision-making. A useful definition has to take into consideration the complex nature of change. TAMI found it necessary to make an interpretation of impact, capable to catch the many different kinds of effects that TA can induce, and which can serve as contributions to a process of change (see the chapter "Towards a Framework for Assessing the Impact of Technology Assessment").

Impact is an induced change with regards to the state of knowledge, opinions held and actions taken by relevant actors in the societal processes on science and technology issues.

3.3
Method and impact

A method has to be seen in the light of its surroundings. A method is a tool, used in a project or activity to serve a certain purpose, framed by the project organisation, the host institution, the actual situation of the issue under scrutiny, and, in broader terms, the political and institutional culture. The meaning of a certain method is in other words defined by the context. But at the same time, a method used in a technology policy analysis intervention makes changes to the context. The state of the issue, the condition for the actors involved, as well as the institutional and political situation may be affected through the TA intervention, and these changes – the impact – are, among other things, a result of the specific method.

3.3.1
As seen from the method

TAMI has developed a description of the relationship between the method and its context, which is described in-depth in the chapter "The Practice of Technology Assessment; Science, Interaction, and Communication". This method interpretation puts weight on the procedural aspects of TA activities, with *scientific analysis*, *interaction* and *communication* as the procedural pillars.

As the meaning of a method is dependent on the context and on the "goals of change" which has been setup for the TA activity, *methods have to be chosen in accordance with an understanding of the problem situation – the context*. Part of this analysis has to take into account the wider settings of the TA institution – the political culture, institutional background and history – which empirically can be seen to have had great influence upon the methodology of TA institutions, in particular upon the composition of the toolbox.

Involving both scientific, interactive and communicative processes, *TA has a broad set of quality criteria to live up to, in order to make assessments that are valid and can be trusted* as a promoter of change. *Scientific quality* includes the proper management of interdisciplinarity, of reviewing processes and professional discourse. *Quality of interaction* involves social as well as procedural fairness, transparency and the quality of arguments. *Quality communication* has to do with flexibility towards the ongoing debate, ensuring relevance with regard to the social, political and scientific reality, and with focusing on target groups, media and strategic alliances.

The choice of methods has a potential impact on the wider political perception of the TA institution and its role. This is a consequence of the fact that methods have a political cultural meaning connected to them. Consequently, the *composition of the toolbox and the choice of method is an important part of institutional strategy*. A method that may contribute to building up trust in one TA institution may have the opposite effect in another institution. On the other hand, the establishment of a TA institution may in itself be a reflection of the need for a change in political culture, in which case not only new fields of method experimentation are opened up, but also the exploration of these fields may be a prerequisite for building up trust. Trust and recognition are important determinants of the general impact of an institution, which leaves the choice of method to be not only an instrumental choice, but also a strategic choice.

Though the method should be chosen to support the goals of the TA activity, the *method is certainly not the only determinant of the impact*. The success of TA projects is dependent upon other internal factors (project management competencies; budget constraints; organisation culture; etc), boundary conditions (institutional setting; links to policy-making; political culture; etc), as well as an infinity of more or less controllable external influencing factors (strategies of other actors; timing of policy-making process; sudden changes in the problem situation; the immediate issues on the media agenda; etc, etc). A method may be chosen and adapted to compensate for as many as possible of the relevant influencing factors, and some methods are stronger in setting the agenda than others, so, a conscious method implementation will certainly lead to greater chances for impact. However, the examples of unforeseen obstacles for impact are many, which to some degree has to do with the fact that no project design can control all influencing factors.

An important part of maximising the impact of TA activities is to *acknowledge the importance of conscious communication strategies connected to the method*. Communication is an embedded part of all knowledge sharing, social learning and policy definition, and as such, it is the backbone of all interaction. The more consciously and professionally the communication inside as well as out of the TA process is handled, the more readily will the actors learn from the process and its results. Communication strategies should not only focus on the outcome, in terms of reports, briefs, follow-up conferences etc, but much more importantly focus on the interaction included in the method, and the spin-off in terms of involvement of actors and networking along the project process. In a broader sense, communication techniques and skills should be regarded as an integrated part of the methodology competencies in a TA organisation.

The methods of TA are generally well described and studied with regard to the range of stakeholder involvement and public consultation methods of participatory TA, of which manuals are often available and a worldwide exchange of experience is common. Different attempts have been made to build up typologies of TA and policy analysis methods, but none of the attempts has succeeded in the establishment of a generally accepted typology. TAMI has found that the lack of a common typology of TA methods, can to a large extent be explained by the fact that a refined picture of traits of different methods does not exist. In particular the different traits of scientific or expert based TA approaches needs to be examined and described with regards to their procedural differences. Further *research is needed to make an in-depth overview of TA methodology, and to suggest a method typology* that connects the method to a set of goals of the TA intervention.

3.3.2
As seen from the impact

TAMI has developed a *typology of impact*, which is described in the chapter "Towards a Framework for Assessing the Impact of Technology Assessment". The typology is build on the assumptions that A) TA can *be about* the technology/issue under scrutiny, the societal aspects, or the policy aspects, B) TA can *lead to* new knowledge, new attitudes/values, or new actions taken, and C) there are a number of *specific roles* that can be played by the TA project inside the framework made up by A) and B).

The idea of roles[16] as an expression of the often delicate and complex changes evolving from a TA activity that involves for example experts, stakeholders, policy-makers, public authorities and NGO's in the same TA process, makes it possible to characterise those kinds of impact that are often referred to as "soft" or process-oriented. However, though delicate, soft and process-oriented, *impact understood as a "role" is a concrete change in the societal situation on the issue at stake.*

TAMI has identified an inventory of 21 roles, which have been seen performed in TA project examples. The list is non-exhaustive, and future case studies may reveal new roles. The introduction of the *concept of roles reveals that TA plays more roles and has more impact than usually appreciated* in the discussion of the impact of TA and other forms of science and technology policy analysis.

An important conclusion of the TAMI project is *that defining impact as a change in decision-making is much too narrow.* This kind of impact makes up only one "corner" of the TAMI typology. It is often questioned, if TA should search this kind of direct impact on decision-making at all. There are examples of direct and targeted impact on decision-making that have been regarded as legitimate – for example in terms of campaign-like TA activities towards decision-making at local or enterprise level. However, *it seems to be in contradiction with the role as advisor to be expected to exert direct influence upon political decision-making*, and even if it is expected, TA lacks the methods, which legitimately can intervene directly into the processes of representative democracy.

A TA project goes through a series of phases (as portrayed in Figure 1 of the "The Practice of Technology Assessment; Science, Interaction, and Communication"), of which some of them are occupied by the execution of the included methods. When impact is regarded as an ongoing process of change during the TA project, which can be identified as specific roles, it becomes apparent that *all project phases are potential platforms for impact.* For example, just the fact that a TA institution takes up an issue (the first phase of a project) may in itself play the role of "setting the agenda". Or, an expert/stakeholder seminar at the end of a desktop research TA project in a phase of synthesizing policy options may play the role of "building bridges" between the actors.

Certain roles are method dependent in the sense that they are typically expressed by certain categories of methods. For example, the role "self-reflecting among actors" is hard to perform in a desktop research activity, and will be facilitated more easily by an interactive method of participation. Further, to play a certain role may be dependent on the other roles that are being played. For example, a change in legislation is often dependent of knowledge and debate. Therefore, there are ties

---

[16] TAMI has developed the concept further from the ideas presented by the EUROPTA Project by Danielle Bütschi and Michael Nentwich: "The role of Participatory Technology Assessment in the Policy-making Process", in: "Participatory Technology Assessment, European Perspectives", Ed. Simon Joss and Sergio Bellucci, Centre for the Study of Democracy, London, 2002. See also www.tekno.dk/europta

between the ex ante choice of "goals" for a TA activity, the expected dynamic relation between "roles", as it is expressed in the project design, and the ex post registration of the roles played by the project. Or in other words, *the TA project goal has to be broken down into the required roles, as part of the project design.* How to ensure that the project performs certain roles has to do with the competencies (process understanding, communication skills) of the project management, with realistic and strategic setting up of expected roles during the project timeline, and with a careful choice of methods.

There are, as stated above, many influential factors outside the control of TA, which may have immense impact on the type, amount and direction of impact that the TA activity can exert. However, in connection to each role, intended to be played by the TA activity, *there are means that can be set in effect to counteract unwanted influence, and to amplify the specific role.* The problem situation always leaves a certain degree of freedom for the actors to change strategies, which leaves it up to the TA intervention to be prepared to be flexible.

It has not been a research aim of TAMI to make in-depth case studies to examine how the roles are played inside the various methods, institutional and political settings, issues, or any other potential factor, which may influence the impact of TA. However, such research is greatly needed, as the key to playing the relevant roles on a sufficient scale lies in the ability to choose the right measures, taking the relevant factors into account. Therefore, *empirically based research on the instruments connected to performing specific roles needs to be set up* in order to build up further efficacy in policy analysis exercises. The institutional and political settings need special research attention as factors that may determine the acceptability of or presuppose the use of certain measures (such as for example citizen participation or direct contact to politicians) and therefore may affect the ability of the project to pursue a certain goal effectively.

3.4
Evaluation of impact

Historically, impact of TA has been difficult to measure. There are *examples* to be found of all kinds of impact, but there is scarce *evidence* of direct impact of TA on decision-making. This is not at all surprising from a political science theoretical point of view, since TA – as any other policy analysis – serves decision-making with knowledge, options, networks and arguments, which are highly valuable raw material for the political process, but which also are difficult to trace after the workings of the political foundry.

The TAMI project has made some major improvements in the understanding of the nature of TA impact, by a) establishing an impact typology that b) involves a set of "roles", which c) can be seen as a relative realisation of the "goals" of the TA activity. It has not been the intention of TAMI to develop an evaluation methodology for TA, but retrospectively, it appears that *the systematic use of "roles" as the expression of impact makes evaluation easier by role-specific evaluation indicators.* Identifying the effects of TA activities still is a difficult task, but the more precisely one knows what to look for, the easier the search gets.

The TAMI partners are convinced that a structured evaluation tool can be set up for any TA project, on the basis of the priorities that are given to the roles that are to be played by the project. Such a tool should differ between the ideal intentions in the project design, the realistic expectations with regard to the scale of impact (level at which certain roles is expected to be played, taking – among other things – the involved method into consideration), and role-specific evaluation criteria, which may be documented with a variety of tools. A possible layout for such *an evaluation structure can be sketched as an inventory of the TA goals, their possible level of expression, and their specific evaluation indicators*:

Goals	Realistic achievements	Evaluation Indicators
Priority list of roles	Adjusted expectations to each role with regard to situation appreciation, project design and influencing factors.	List of indicators for each role: – Hard indicators – e.g. opinion surveys, media coverage, official proceedings – Soft Indicators – e.g. interviews, personal judgements

The general scheme for an adequate evaluation methodology that respects equally the scientific, interactive and communicational nature of TA seems to have come much closer with the introduction of the TAMI typology. Such an evaluation tool would be extremely valuable for cross-national comparison of the outcomes of TA, foresight and other science and technology policy-analysis functions. However, *further research to develop this tool into a general benchmarking technique is needed*, and would be highly relevant.

3.5
The institutional setting

TAMI has seen the institutional setting as a specific "influencing factor" on the types and level of impact that can be expected from TA activities. *Institutional boundary conditions may affect all phases and aspects of the TA process*, such as:

- Problem selection and definition. For example through presumptions and values embedded in the organisation and its surroundings.
- Actor relations and involvement – close or loose connection to important actors; ability to connect to actors as needed.
- Toolbox composition and method choice – which may be affected by the institutional relation to certain communities (scientific institutions; trade or employer organisations; parliaments etc.).
- Project management and interaction. May be affected by the expectations to the function of the TA staff – are they to act as researchers, process managers, policy-makers or civil servants?
- Communication. How freely can the institutional choices, the TA process, the involved knowledge, values and decision-making, or the conclusions and their consequences be communicated?

Looking at the worldwide institutional landscape of those TA institutions, dealing with parliamentary TA[17], *three models of institutional settings are visible*, all of them imposing different influence on the methodology and impact of the institutions. The three models may be called the scientific institute, the parliament office, and the public institution.

The scientific institute is characterised with a more or less formalised link to academia, whether this is in the form of connection to a university, to the research councils or to an academy of science – historically, such setups have been seen. This type of institution typically has its strength among those types of impact that have to do with the research, analysis and synthesis of knowledge – the cognitive dimension of TA. The limitations can be found among those roles that can be regarded as being in conflict with the role as scientist. For example, it can be difficult for these institutions to take an active role in setting the public agenda, to involve itself in interactive normative clarification, or to set up political processes. Historically, at least for parliamentary TA institutions, this model often seems to be an intermediate state towards an institutional setup, which has closer links to policy-making and a toolbox that goes beyond the scientific analysis.

The parliament office serves parliamentarians directly, and is set up as an internal office of the parliamentary institution. The strength of this type of setup seems to be connected to the clear focus on the needs of MPs, which favours impacts on the knowledge and policy dimensions. The main limitation is the other side of that coin – a low ability to set up TA as a learning process directly involving or put in to the hands of stakeholders and the general public, which makes impacts of opinion-forming and debate less probable. In terms of methods, this means that these institutions mostly focus on scientific consultation – often in connection with hearing-processes involving stakeholders –, and have lesser focus on methods that support the social and political processes outside the parliament.

The public institution may be set up in connection to the parliament or the government, or other constructions. It has a high degree of self-governing competencies, obligations towards the societal discourse as such. Parliamentary TA institutes set up as public institutions also have obligations towards the parliament, and most often a direct link to a parliamentary committee. The autonomy to take up issues and pick methodology as the institution finds it most relevant is an important feature of this organisational setup and this is reflected in the use of a broad toolbox, which potentially can result in any of the impacts/roles in the TAMI typology. The downside of this setup, however, is mostly to be found in the low involvement of politicians in the internal process of the institution, which may hinder the call for TA as part of the parliamentary decision-making process.

Though such models can be identified and can be seen to influence the methodology and the impact, *the choice of roles has as much to do with institutional strate-*

---

[17] We focus here on the parliamentary TA institutions, as they have a common aim of directly supporting the policy-making process. Other TA institutions may resemble these with regard to institutional setting and analytical process, but may not have the same kinds of incentives that are involved in PTA.

gies as it has with formal boundaries. Inside the models, the aims of the institutions are always subject to negotiation, which means that the right strategies may open up new methodological fields to explore, and, subsequently, for new and more roles to be played. For example, the *scientific institution* may open up for interactive methods involving stakeholders or citizens, in order to supplement the scientific analysis with debate; the *parliamentary office* may decide to take in citizen participation in order to serve the parliamentarians with in-depth knowledge about public opinion; and the *public institution* may compensate for the weak links to the parliament by going into cooperation with the parliamentary committees directly on an ad hoc basis. An important feature of TA seems to be that such strategies are real, and there *is a tendency for all types of TA institutions towards expanding the toolbox and playing a more diversified set of roles.* These points are supported by the fact that across the different models and the many different kinds of TA being performed in our societies, there is a common understanding of the rationales and the mission of TA activities.

From the assumption that the establishment of TA offices is an expression of a general need for more conscious technology policy-making, it follows that such offices should be present, wherever knowledge, opinion forming and action is necessary. This may be among MPs, but it may be among stakeholder, experts or the general public as well. In its consequence, this means that the strongest TA is to be found when certain TA goals can be unrestrictedly pursued with regard to method, target groups, communication efforts, etc. Because of that, *it is of high importance to study how and with which results, institutions can overcome traditional barriers of political/institutional setting* in order to maximise its impact. Such research would give important knowledge about the "open but unexploited fields" of roles to be played by different settings of TA.

3.6
Communication and process

In parallel with the general movement towards expansion of the institutional toolboxes, an *increasing awareness of the importance of communication as an integrated part of the TA methodology* can be seen. This awareness means that besides the traditional methods of publication, an active approach to communication becomes an integrated part of the TA institutional policies.

The communication between the project activities and the actor networks is crucial for the total impact of TA, understood as the set of roles that are played by a TA activity. Some roles may be possible to play with a very simple communication effort, but *many roles are dependent on an active communication strategy in the project, supported by an institutional communication policy.* If TA is to reach the relevant actors, it will have to be able to compete on the crowded information market, since TA is not the only one to serve the public and political debate.

A *communication policy* may for example imply staff competence, clear rules on competencies to initiate communication activities, a set of ongoing communication channels (web-site, e-mail newsletters, magazines, report series etc.), and project specific communication plans. Communication integrates with all levels of the institutional life, as *a communication policy will be an institutional orientation, a professional skill, and an integrated part of the toolbox.*

A *communication strategy* needs to be tailored to the specific situation of a TA activity. This may for example involve the definition of target groups, and the specification of communication channels to reach them. A strategic approach to communication will, however, also imply a strategic composition of the TA methodology, since *the process of TA in itself is a communicational tool (to participants, audience, organisations, steering committee etc.) with great potential impact.* The choice of method depicts the nature and level of involvement of the actor networks, and therefore is an important decision with regards to strategic communication. Correspondingly, the framing of the issue to a large extent defines the relevance of different dialogue partners in the activity, and therefore the possibility for communication through the process, which makes strategic communication an important part of the design of projects in a TA institution.

The learning processes connected to the roles of TA may be supported by targeted communication efforts in many ways:

– Communication channels for effective "one-way communication" may be built up to support the transfer of knowledge and options from the TA activity.
– The TA project can establish channels of communication with the actors in order to ensure input and learn from the actors, and in order to establish bedrock for the communication of project results.

Processes may be set up to activate the dialogue between the actors, in order to support the mutual learning, the normative clarification and the preparation of action among politicians, experts, stakeholder or individual citizens.

Shaping the long-term TA involvement on a science and technology policy topic as a process of iterative learning cycles can support social learning, as it gives room for actors to develop their standpoints through time, and for the TA institution to flexibly adjust to the changing social understanding among the participants.

3.7
The trans-national perspective

Science and technology issues are by nature involving all levels of governance in our society – from the daily decisions made by the individual citizen, to enterprise strategies, or to regional, national, trans-national and global decision-making. Therefore, TA can be seen as a universal function for all governing levels, which can help any actor to create knowledge, form opinions or to make policies.

Though TAMI has explored methods and impacts, mainly on basis of national or sub-national experience, it was an intention of TAMI to synthesize conclusions on the trans-national, European level. This last section of our conclusion paper should therefore be devoted to the level that has brought TAMI together.

Regional science and technology policy-making has come into focus during the last decade. Globalisation of science, technology development, production, market places and labour markets has forced regions to develop their own awareness of the available resources and their potential, and to build up strategic visioning. Alliances across borderlines, in terms of foresight activities and regional development plans involving trans-national cooperation can be seen all over the world, and not the least in Europe. Technology policy analysis of course play an important role in this development, and *activities in the recent years indicate that national, regional and transnational foresight and strategic analysis may become a major task for TA in the future*, since TA has the necessary interactive, methodological, analytical approaches and strategic aims in place.

The challenges of governance of science and technology issues at European level are fully documented and politically acknowledged. Also acknowledged is the fact, that much of the problems of governance may very well find their solution in effective knowledge sharing throughout the EU society, in a widely embracing strategy of involvement and debate, and in a more open process of decision-making. The TAMI impact typology has been made in order to break these general aims down into a set of specific roles that can be performed by TA. Looking at this typology through the lenses of the state of EU S&T policy, it clearly appears that *the TAMI impact typology encounters the governance challenges of EU science and technology policy*.

This is not surprising, since historically, parliamentary TA was established in order to meet the same challenges at a national level. The nearby question then is: *Is it possible and relevant to adapt, transfer and use the broad toolbox of TA to the trans-national level? From the TA practitioners' point of view of TAMI, the answer is positive*:

- Most methods of TA have been shown to be transferable between nations worldwide. Trans-national events should be big events, but basically they would need a European problem analysis, and could often be built as coordinated parallel national events with a European synthesis phase.
- Methods of TA can be performed by the trained professional community of TA, at European level in accordance with the principles of open coordination.
- Cross-national TA in the pursuit of "roles", as described by TAMI, would be extremely relevant for the proper governance of European science and technology issues, such as transportation, energy policy, bioethics, intellectual property rights, consumer regulations, the brain-drain dilemmas, etc.
- TA activities at the trans-national level makes it possible to transfer competence and capacity to regions and nations inside as well as outside Europe. It would be an important mission in itself for Europe to pave the way for permanent establishment of TA functions and institutions.
- Trans-national TA activities, aimed at analysing and finding policies towards societal conflicts on science and technology issues are, however, confronted with some important questions that need to be answered methodologically. Such "scaling-up" of problems are often debated, but there is reason to believe that they can be solved by proper project designs:
- The lack of a "European public sphere" is often mentioned as a barrier for initiating debates on the trans-national level. However, the lack of a European public does not mean that European issues cannot be treated in the public room. Focusing on nations as the operative level for trans-national TA seems to be the answer.

- Cultural differences in communication and debate on political issues are other challenges for trans-national TA. The answer may be to explicitly make use of a bi-focal approach in which as well as the method,the content could vary from nation to nation, inside certain controlled limits.
- Language barriers are obvious obstacles for European public dialogue. However, national TA has overcome this barrier in several cases.

It is the impression of TAMI that *the specific challenges of performing TA at the trans-national level are possible to solve.*

There is an *increasing call for the methodology of TA to be implemented in a European trans-national context in order to harvest the impacts and societal roles of TA.* In technical terms, there seems to be nothing hindering trans-national coordinated TA activities. Further, the TA community represent a unique network of mainly national competence that is ready to go into coordinated trans-national activities. The immediate need does not seem to be the development, research in or further case studies on the already known methods. At the time being, there seems to be a much more *urgent need for actual projects at a European level.*

Part II – Supplementary Papers

Shaping the Impact: the Institutional Context of Technology Assessment[*]

Laura Cruz-Castro and Luis Sanz-Menéndez[$]

1
Introduction

This paper addresses the relationship between scientific knowledge and politics, and more specifically the role of knowledge and information about highly technical issues in politics. The connection between science and politics, the relationship between government (the executive) and science (Smith 1992) has been studied at length, both from the viewpoint of science as a system that demands funds and resources (Cozzens and Woodhouse 1995), and of the role that scientists and experts play as advisors in the political process (Barker and Peters 1993; Bimber and Guston 1995), to presidents (Bromley 1995), bureaucrats (Jasanoff 1990) and even to the judicial system (Jasanoff 1995). Some steps have also been taken to gauge the impact of scientific advice (The IPTS Report 2001; The IPTS Report 2003).

Fewer studies have focused on the relationship between science and the legislature, or between scientific knowledge and parliamentary decisions, with the exception of the literature on the US Congress, and specifically the Office of Technology Assessment (OTA) (Gibbons and Gwin 1988; Bimber 1996; Bimber and Guston 1997). However, it is also true that there is a steadily growing array of literature on the provision of scientific and technical (S&T) advice to parliaments in Europe (Vig and Paschen 2000), and on the demand and use of technical information by the US state legislatures (Jones, Guston and Branscomb 1996).

The aim of this paper is twofold: First of all, to analyse the emergence, nature and way in which parliamentary "scientific advice" activities are carried out in different european countries. In particular, the paper will concentrate on examples where advice is provided through a specific practice called "Technology Assessment (TA)" and through organizational forms of production of this knowledge that, albeit differing, are known as Parliamentary Offices of Technology Assessment (POTA), and provide a service to parliament, either as their only customer or as a special customer. Secondly, to outline the similarities and the differences in the organization of these POTA and, from an institutionalist perspective, to analyse and explain the diverse impacts of these organizations' S&T advice or TA activities.

---

[*] We thank TAMI colleagues for their comments, special thanks to Lars Kluver, Jan Staman, Leo Hennen and David Cope for their suggestions. We also thank the European Commission (HPV1-CT-2001-60043) and the Spanish Ministry of Science and Technology (SEC-99-0829-C02-01) for their funding.

[$] Corresponding author. Tel: +34-915 219 160 (ext. # 105); fax: +34-915 218 103; email: Lsanz@iesam.csic.es.

Analysis of this specific knowledge production instrument known as technology assessment (TA) has been heavily dominated by normative approaches, which have even been characterized as various paradigms of TA (van Eijndhoven 1997), and methodological approaches (Porter 1980; Kuhlman et al. 1999; VV.AA. 2000). This has undermined its value as a practice associated to the design of science and technology policies (Smits, Leyten and Den Hertog 1995) or to the forging of stronger ties between science and society (Rip, Misa and Schot 1996).

As for TA's connection with politics, many authors have analysed the political process that led to the creation of the OTA, the first parliamentary technology assessment organization, its subsequent operation and the causes of its demise (Bimber 1996); others papers have focused on the influence that social and cultural factors have had on the dominant form adopted by POTAs in several european countries (Petermann 2000) or on the institutional aspects of their establishment (Vig 2000), but most of the studies describe how the different POTAs are run (Paschen 2000; Kluver 2000; Laurent 2000; Norton 2000; etc.)

In our opinion, the nature of these unique information production practices and their emergence are phenomena that can only be understood within their institutional context. Our contribution places the institutions, the rules of the political game and the types of incentives that TA organizations face, at the heart of the explanation for their adaptation, consolidation and differential impact, pointing out the contradictions that sometimes exist between the institutional arrangements and the relative impact of the TA produced by the organizations linked to parliaments. We argue that the *institutional arrangements that govern POTAs* are key factors for explaining the depth and extension of the impact of their TA activities; therefore, if one wants to understand TA's impact in the political and social process, one must first understand and characterize the institutional and political context in which it takes place.

Our analysis is constructed upon the comparative method (Collier 1993), albeit in circumstances determined by a small number of cases (*n* small). The argument will be contrasted empirically with a set of cases from several countries that have created "parliamentary scientific advice" institutions by analysing their institutional arrangements, as well as the type of TA that they perform and its impact. The information has been collected from questionnaires and interviews conducted with the people in charge of these organizations. The impact is gauged on the basis of the role description and the typology of possible impacts drawn up by the TAMI team.

In section two, and since history has a bearing on understanding the present (Rose and Davies 1994), we take a brief look at the emergence and consolidation of TA as a practice that produces information about S&T matters, oriented and linked to the political process, specifically to parliament. In order to explain the emergence of TA, in section three we put forward a theory regarding the use of scientific advice and information in parliament and in policy making in general. To this end, we adopt a hypothesis that complements the traditional rationalist model of "information" as input in the decision-making process, and that also regards information as a mechanism for legitimising decisions and as an instrument in power struggles. Section three puts forward an analytical structure for understanding how TA has emerged in different degrees as a practice with its own identity in several countries and, above all, for explaining the reasons why TA is or is not adopted in its parliamentary form, i.e., linked to the political process, and what types of arrangements are established. Section four analyses the institutional arrangements and organizational structures in which TA occurs in the parliaments of some european countries. We define and characterize the two models or types of arrangements governing TA as the practice of producing information for the political process: the instrumental model and the discursive model. Section five explains the types of impacts and their relationship with the diverse institutional arrangements existing in the POTAs. Some empirical evidence is then used to establish general descriptive and normative hypotheses. Lastly, section six, presents some conclusions.

2
The emergence of TA as a specific cognitive and information practice

In this section we characterize the concept of technology assessment (TA), that unique variety of production of information or "policy advice" related to S&T matters. To that end we take a historical approach, describing the emergence and creation of TA as an information-production practice, with its own identity and special ties to the political process, as the outcome of a specific juncture in the United States, in which it attained a high degree of institutionalisation.

Broadly speaking, TA has been defined as the production of information about the possible consequences of S&T developments to improve public policies. More specifically, we regard TA as a type of information about highly scientific or technological issues that tends to come in written form, in documents or reports, and that is developed to "improve the information for policy-making"; in a few evolved forms, TA also seeks to foster debate, public understanding or acceptance of the impacts of science and technology from a neutral, non partisan position, always using scientific information of the best possible quality[1].

In the United States, the ever growing importance of science, not only because it demands resources from our societies, but also as a core element of political decision-making, gained relevance in 1957, after Eisenhower created the post of the President's Science Advisor and the "President's Science Advisory Committee" (PSAC) (Killian 1977).

By the mid-Sixties, US society harboured a growing concern about the negative side effects, in environmental or health risk terms, of the development of certain technologies (for example, nuclear technology or pesticides, which won their discoverers the Nobel Prize of Chemistry). This prompted a movement to reassess technological developments with a literary best seller (of Rachel Carson's book *Silent Spring* (1962)) as a flag.

Civil society, especially the scientific community, sponsored the development and production of information, based on the best scientific evidence available, about the "polemical" aspects of scientific and development technological. The goal was to offer policy makers insight that would help them to make decisions

---

[1]   TA has been applied frequently in areas such as transportation, energy policy, bioethics, GMOs, consumer regulations, waste policy, environmental policy, etc. See, for example, the history of the OTA in its reports (OTA, 1996).

about impact-relevant projects. Broadly speaking, the analyses, originally sponsored by the research facilities, were designed to provide advice about scientifically relevant issues and, in short, to influence political decisions. What one might call "grass roots TA", or what Van Eijndhoven (1997) labelled "classical TA", developed in the Sixties, especially in wealthy countries such as the United States, and peaked when it expanded its boundaries and merged with politics. In October 1972, the United States Congress decided to set up a specialized internal agency, the *Office of Technology Assessment* (OTA), stating that "the basic function of the Office shall be to provide early indications of the beneficial and adverse impact of the applications of technology and to develop other coordinant information which may assist Congress" (92nd Congress, 13 October 1972, Public Law 92–484) (quoted by Holt 1977).

The creation of the OTA should be seen in the light of the effect of those concerns about technology's negative impacts, or the technological optimism that has been the hallmark of US bureaucracy; yet it is also essential to place it in the right political context. The United States is a political system marked by a "divided government". It should also be seen in the light of the effort that Congress made, at the start of the Seventies, to bolster its position vis-à-vis the executive (Bimber 1996). Thus, the two key factors behind that unique event were: the aim to reinforce the scientific and technical information available to parliamentary decision-makers with non-partisan, unbiased information, of the best scientific quality available and, secondly, the aim to bolster Congress' power vis-à-vis the executive, an institutional effort that is also to be observed in the creation or the reinforcement in those years of other Congressional agencies (Sundquist 1981). The OTA, whose creation was initially sponsored by the Democrats and triggered certain distrust among the Republicans, finally managed to earn itself a reputation as an agency that provided non-partisan advice, in order to improve decision-making within Congress.

The OTA and its activities managed to afford *"technology assessment"* an identity, as information based on scientific and technical knowledge, but different from scientific knowledge in the strictest sense and from the information produced by Washington's think tanks or interest groups, and geared towards offering neutral advice to Congressional policy-makers. Further on we look at the explanatory significance of the impact caused by the fact that an internal, instrumental model for producing information and advice on scientific matters for public policy, such as the OTA, arose within the institutional framework of a divided political system; one with a clear separation of powers between the executive and the legislature, in which legislative power has been traditionally decentralized in the committees and sub-committees (Polsby 1968; Cooper and Brady 1981; Polsby and Schickler 2002). It is against this political background of decentralization where an internal instrumental model has the strongest capacity to influence political decisions, even in the absence of strong external ties. However, the same circumstances that allowed the OTA to enjoy extraordinary resources, and therefore facilitated its impact, were no guarantee for its survival when the political conditions changed radically in 1995.

Therefore the first successful example of TA as a practice in a political context can be explained by circumstantial and local reasons. However, this activity, which took place with political institutionality in the United States, and its experiences contributed to afford TA an international identity. From that moment on, the OTA and TA model became available for rational imitation (Hedström 1998), international policy transfer (Wolman 1992), or inspiration in the context of drawing lessons for policy (Rose, 1991). So to understand why the same type of TA was developed and adopted in other countries' political systems, as well as thinking about local preconditions, one must pay attention to other issues such as the international diffusion of models and policy ideas, or the emergence of an epistemic community around TA as a practice.

The next section explains how the diffusion, imitation of, and inspiration in the OTA model were influential factors in the development of parliamentary TA in Europe; however, it seems evident that today TA is not a significant source of information in guiding technology policy throughout the world. The reason for this perhaps lies in the general nature of its method of work, subject to the political processes of conflict and agreement; or perhaps in the fact that it has not incorporated or constructed a "policy paradigm" (Heclo 1974) or has been unable to form an "epistemic community" (Haas 1992) in which the producers of policy ideas and policy makers share the same model and goals; or, perhaps, in the fact that there are countless sources of information that compete for the "limited attention" of political organizations' decision-makers.

3
The diffusion/adoption of TA in different national contexts

If the traceable origins of both the grass roots and institutionalised versions of TA are to be found in the United States, then two questions remain to be answered: first of all, why has TA developed as a type of unique S&T information, with its own identity, in some societies, but not in others? Secondly, why was TA adopted in its parliamentary form? Why were parliamentary offices of TA set up in some European countries?

This section puts forward an analytical structure for understanding the emergence of TA as a specific practice of information production on S&T policy matters in the different European countries. It also outlines the factors that explain why some countries adopt the parliamentary form of TA while others do not, and the types of arrangements that are established.

To answer these questions about TA, we must distinguish between two levels of analysis: the analysis of the emergence of TA in general produced by different groups, and the analysis of TA's ties with politics. First of all we will analyse the general conditions in which the supply (the producers) of this type of information arises and, then examine the social construction of demand from the political system and its institutionalisation.

In the cases we examine, there seems to be a sequence in which the supply of TA or a form of quasi-TA arises first and is then followed in time by demand. This is in a context of political processes that draw on other countries' experiences and recently the activism of the European Parliamentary Technology Assessment (EPTA) network seems to be a key factor.

Until now, no clear answer has been found to the first question, because most of the case studies conducted have a selection bias and only analyse countries where these organizations have indeed developed, and/or also in which TA has been linked to the parliamentary scientific advisory process. One might suppose that the US precedent spread, leading others to imitate and draw upon the OTA's experience, yet the almost total absence of TA in some countries seems to indicate that this information production practice only emerges if certain prerequisites are met.

After analysing the cases in which TA has developed in general and where the groups that produce TA have flourished, we can put forward a general explanation associated to the importance of local preconditions. Three macro social factors serve to explain the relative abundance of this type of practice, and the consolidation of groups or organizations in the general panorama of socio-politically oriented "scientific and advisory" institutions: first of all, the country's level or degree of S&T development, measured by the percentage spending relative to GDP or by the number of researchers per person in employment; secondly, the existence of a problem of "public perception of science" (whether negative or positive) in each of the countries, and a certain balance between the "hope and fear" that the technological developments generate; lastly, in some countries the civic and participatory nature of democracy matters, insofar as TA is generally a grass roots movement.

In short, one could say that this unique type of information, TA, can only exist if there is already a certain level of S&T development in those societies, where the scientific community is concerned about the social responsibility of science, and also where the new social movements such as environmentalism have been politically relevant. TA is the normal outcome of the existence of a "rich" scientific system, where research facilities have surplus capacities and abundant manpower because, generally speaking, the first people to practice TA are the very same scientists who, for different reasons, try to influence the course of political decisions. Therefore, TA can emerge – and as a matter of fact has emerged – regardless of the context within which it is used; in theory it can emerge even without any explicit demand from the political system. However, experience shows that the density of TA and S&T policy analysis organisations is a good predictor of the emergence of the political system's demand[2].

With low levels of S&T development, there is little opportunity to make decisions about scientific matters. And without democracy, which is also closely associated with economic growth (Przeworski et al. 2000), and without a civic culture, the road to decisions about public policies is closed to practices such as TA.

Hence TA is a special type of "policy analysis and advice" produced by scientists, or heavily influenced by scientific method; yet to gain acceptance it must compete – and try to coexist – with many other sources of information that politicians, governments, bureaucrats or parliaments use. As a practice mainly of the academic community, or of social movements, TA has had to compete with other information, because there are many other sources that produce information about "the possible outcomes of technological options": the experts who work in or for the government, regulatory agencies, bureaucrats, think tanks, interest groups, etc.

In our countries, there are plenty of sources of S&T related information that can inform public decisions, and that seek to influence them; one might say that the different organizations that produce information and "advice" on political matters, many of which relate to S&T, compete with one another, and that when one of them

---

[2] A brief look at the database of TA institutions in the world seems to confirm this argument.

emerges, there is a greater likelihood that more will appear, though perhaps with less of a likelihood of only one of them being used by socio-political actors.

One form of S&T information production quite similar to TA – and perhaps with shared roots –, but that has normative powers and the authority to impose sanctions is the one that has emerged in the two last decades in the "regulatory agencies". These are unquestionably specialized competitors of other TA organizations; this "variety of TA" has developed under the executive, especially in the technology related regulatory agencies, albeit sometimes with institutional arrangements that tie them to the legislature, as occurs with the Spanish Nuclear Safety Council; however, in these cases, even if the production of S&T information is guaranteed to have an impact, its quality or independence has varied greatly. The missions of the regulatory agencies have been clearly defined and specialized, as in the case of those responsible for nuclear safety, evaluating drugs or environmental technologies and, more recently, health technologies. These regulatory agencies have built up such highly specialized S&T capabilities that, in most of the countries, they enjoy a quasi-monopoly in their respective fields over the knowledge about the aspects or problems associated to those technologies. In fact, regulatory agencies could be said to have spread as a result of the pressure exerted by social actors and the technical "rationality" being imposed in the settling of conflicts.

If the development of regulatory agencies with very strong S&T powers in traditional TA fields such as the nuclear industry, the environment, etc, that report to the executive, has not hindered the growth of demand for TA by the parliaments, perhaps it is because parliaments have helped to institutionalise TA and afford it an identity as an information production practice and a generalist methodology. Furthermore, it is worth assessing whether the development of these regulatory agencies is likely to threaten the survival of POTAs.

So it seems that there is something about TA, its method, and how its identity has been built that makes it especially prone to being connected with parliaments, to the institutionalisation of its relations with the legislature. That is why the second part of this section analyses the development and institutionalisation of several varieties of TA linked to the political system, and more specifically to parliaments, not just with a symbolic or rhetoric, but with a substantive nature in those countries that had met the prerequisites. The concepts that are used to argue this bond (demand) have to do with the legitimation of decisions, with power, and more specifically with the distribution of power.

As we have said, TA, insofar as it is an information production and awareness practice, became established and acquired its own identity in association to the political system. Perhaps this was due to the impact of the experience of the US Congress OTA and its subsequent international diffusion, and with the intention of influencing public policies with a high S&T content. Understanding this bond involves adding an alternative hypothesis to the traditional rationalist model, which regards "information" as input in the decision-making process, so that information is also regarded as a decision-legitimising mechanism. Organization theorists have long known that organizations have and demand a lot more information than is actually used in decision-making processes. That is why they say that "having the information" is quite often a legitimation mechanism, rather than an aid for making decisions (Feldman and March 1981). Such decisions are usually made applying rules different to those imagined for decision-makers endowed with "Olympic rationality" (Simon 1983).

In this context, the volume of information from different sources, and sometimes linked to specific groups and interests, means that decision-makers can be faced with a lack of legitimacy, and thus claim that they do not have any neutral, independent information. To understand this, one should remember that the TA produced by the organizations linked to the parliaments in the different countries, is only one of the "types of information products" related to decision-making that is available. But what is special about it is that this information is produced institutionally in the parliament, at the request of parliamentary bodies, and that parliamentarians may have influenced when and how it was produced, giving it a "owner" identity.

With such a proliferation of alternatives sources of policy analysis production, the recipients may become saturated, and the mass availability of information from interest groups can also lead to politicians wishing to receive "neutral" information, a wish that can be satisfied by an independent organisation. As true as it may be that, if the impact is the power to influence, that power is inversely proportional to the existence of alternatives sources for achieving the same information resource (Emerson 1962). At the same time, if the need for neutral information is essential in order for politicians to claim legitimacy, then it is just as relevant to argue that decision-makers' attention capacity is limited, and that any channel of direct access to them, with the right timing, is a major organizational asset. This latter case includes the POTAs.

What is specific about parliament-linked TA, as part of its identity, are the "neutral", "independent" and "non partisan" values of the information produced, values built on traditional Mertonian scientific exceptionalism. It is in this context that the claim regarding the "scientific" nature, due both to the method and its significance, of the information produced through TA, becomes all the more relevant, because it implies the use of the legitimacy derived from the scientific knowledge in the political and parliamentary process. Furthermore, despite the legitimising functionality of the information, parliamentarians and experts must build up a relationship based on trust.

Having already found a functional explanation, namely the legitimacy, for the connection between TA, and other types of information, with the political process, another question that remains to be answered relates to the emergence of POTAs in Europe: are they the result of the international diffusion of that political innovation, based on imitation or on any other concept? Or alternately, can POTAs be said to have emerged and emerge as the result of socio-political preconditions?[3]

Looking at the dates on which the european POTAs were created (and actually began operating), one finds a certain process of "international diffusion" or imitation of this US local innovation, namely the OTA. Innovation diffusion processes have had a long tradition of study (Coleman et al. 1957; Coleman 1966; Rogers 1962/95); scholars have studied the international diffusion of organizational models and norms (Meyer and Hannan 1979; Meyer et al. 1992; Finnemore 1996), the diffusion of political innovations within the political system (Polsby 1984; Walker

---

[3]   The problem has already been addressed in literature from the methodological viewpoint and is known as Galton's problem.

1969; Berry and Berry 1990) or on an international scale (Wolman 1992; Rose 1991; Majone 1991).

Rose (1991) spoke of 5 different ways of drawing lessons from other countries' experiences, albeit while referring to public policy programmes: 1) copying; 2) emulation; 3) hybridisation; 4) synthesis; 5) inspiration. This last model often occurs when political decision-makers travel and, upon seeing a familiar problem in an unfamiliar environment, they expand the ideas of what is possible, etc. There is little doubt that the OTA and other subsequent experiences, thanks to the information exchange mechanisms in place between the legislatures (for example, the Inter-Parliamentary Unions, etc.) were a source of inspiration for tackling the problems identified by the European Parliaments. By resemblance, they shed light on easier paths forward thanks to the solutions that others had adopted beforehand. In the european case, apart from national activism, one cannot forget the European Commission's initiatives, such as the FAST programme, which at the start of the 80's contributed to open up a Europe-wide debate, in the context of the European Technology Assessment Conferences, which began in 1987 in Netherlands (Smits 1987).

Yet while the simple diffusion model may well be valid, the fact is that other preconditions are necessary for the diffusion of political innovations. We have already said that socio-economic preconditions explain first the level of development of TA in general. However, despite being necessary, these conditions do not suffice for TA to be linked to the political process in parliaments. There also have to be political entrepreneurs (Schneider, Teske and Miltrom 1995) willing to push forward the initiative in a context in which either they want to strengthen a parliament's position vis-à-vis the executive, or else foster participatory models that involve a better informed citizenry.

According to Kingdon (1984/95), the policy process is formed by 3 different streams: the problem stream (there is a problem involving information or asymmetry of power), the solution stream (here the existence of models of other countries from which to draw inspiration is crucial) and the stream of the political juncture that prompts political entrepreneurs to match problems with solutions. In other words, the political entrepreneurs, from the institutions, "will bring the solution", in the form of POTAs, for the problems that they perceive to exist in parliaments.

An analysis of the individual cases shows that political entrepreneurs were essential in the institutionalisation of POTAs, because they managed to overcome the reluctance of the Parliaments and their majorities, most of whom were little inclined to the experiment, given the nature of continental political-parliamentarian systems. In short, promoting this type of Offices in European Parliaments should be accepted institutionally as a mechanism for "reinforcing" a parliament's position with respect to the executive, with respect to the government. The creation of these organizations was only politically feasible if there was a consensus. The POTAs developed and became consolidated in Europe, although under no circumstances on the same scale as the OTA, as was to be expected in Parliaments that lacked the US Congress' enormous powers. The nature of the european political system, in which parliaments serve to support the governments, meant that most POTAs had to "be authorised" by the latter (Denmark was the exception because the opposition majority in the parliament imposed its opinion upon a minority government) and, therefore, may have been a "concession" or an exchange with others, or a gesture of commitment to the democratic institutions.

In this section we have shown that the development of TA supply and the political construction of demand are two different processes. When TA began to form part of European Parliaments, the OTA's experience began to become known internationally, but only when some political entrepreneurs took it upon themselves to combine the "problems" identified with a parliaments overdependence on the bureaucrats', government's and lobbies' information, with solutions adopted in other environments. Indeed, one of the reasons mentioned most often for creating these POTA is not so much to sift the executive's information, but rather as a means of defence against the avalanche of information from the lobbies. That is why the POTAs' success and survival hinges on their capacity to respond to the incentives of their customers and others who request their information appropriately, in an institutional manner, building a relationship of trust with legislators that serves the key goal of legitimising decisions.

4   Institutional arrangements for TA in parliaments: dominant models and adaptive strategies

This section outlines the institutional arrangements that connect TA with politics and specifically with the parliaments in some european countries. The analysis of institutional arrangements has traditionally sought to distinguish between: a) the conditions in existence when these institutions are created (at whose initiative, in what circumstances, how, etc.), that let the political entrepreneurs play an active role, and in which the constitutional rules are established and b) the effects of these constitutional arrangements on the incentives and opportunities structure to which that actors respond during normal operation of the institutions, and as a result of which they develop adaptive strategies that maximise their survival and influence. In line with this distinction, in this section first we will analyse the origin of POTAs in Europe, then analyse two predominant organizational models, and lastly compare the importance of two fundamental dynamics in the adaptation processes and impact of these organizations in institutional contexts marked by the centralization of political power: on the one hand, the development by parliaments of a sense of ownership of the POTAs' information that gives the latter the edge over other producers of S&T related information, and secondly, the construction of support coalitions by these POTAs.

4.1
Origins

Following the expansion of the idea to provide parliament with S&T advisory functions, and after several failed attempts to imitate the american model – for example, in the German Parliament at the end of the Seventies and start of the Eighties –, from the mid-Eighties onwards, Europe began to witness certain successful initiatives. These were sponsored either by a government, on its own or on a parliament's initiative, or by a parliament itself, to set up different mechanisms designed to provide parliament with an S&T advisory service.

It seems that the decision to start up these S&T advisory functions, devised either solely to support a parliament or with an additional mission, and the adoption of the constitutional model, essentially have to do with the nature of the country's political system. That is the type of specific government-parliament relations, and above all with the civic traditions of civil society and how open the political elites are. The volume of means and resources with which the support mechanism is endowed, or in other words, its organizational capacities to carry out the TA mission, unquestionably mirrors a parliament's desire to assert itself (especially if it is an exclusive service) vis-à-vis the executive, a circumstance that existed in the case of the OTA, which at one time had a 200 people staff. As we have said already, the presence of political entrepreneurs who act as catalysts in the right circumstances is essential to success.

In most of the cases studied, it was parliament who took the "initiative", although in a few cases it was the government who provided the solution (see table 1). In France, the United Kingdom, Germany, the European Parliament and even Denmark, it was the parliament who rubber-stamped the initiative and supplied the resources. In France, the early creation of the OPECST seems to have been a by-product of the demands for the democratisation of french society that spurred the left wing's electoral success and the Mitterand Presidency (Laurent 2000). As for the UK Parliament, the POST arose after several MPs sponsored and demanded the creation of a TA office, which was private during the trial phase and subsequently institutionalised within parliament (Norton 2000). In the case of the European Parliament, the demands made by a powerful Standing Committee were resolved internally by setting up a Parliamentary support unit within the European bureaucracy (Holdsworth 2000). In the Danish Parliament, it was the majority opposition that forced the Government to create it (Kluver 2000). In these cases, and in others such as Germany (Paschen 2000) or Flanders, broadly speaking there was a favourable political juncture that enabled the respective parliaments to bolster their position with respect to the government or technocrats. But in our opinion the core variable is the sense of identity and reliability of one's own source of information production that this model of parliamentary support created to compete not only with the government sources of information, but above all with the lobbies' sources.

In other cases the initiative, or at least the solution to the provision of the service, stemmed from government. The cases of Netherlands (van Eijndhovern 2000) and Switzerland represent a way of responding either to the demands of Parliament, which for constitutional reasons cannot create any new body (the case of Denmark), or to parliamentarians' demands for information and advice on S&T matters. Although, in Switzerland's case the mechanism was created upon a joint initiative between the government and the parliament.

Given the nature of european political systems and the relationship between the executive and the legislature, in both cases the creation of the POTAs entailed the majority government (except in the said cases of minority governments such as Denmark) accepting or tolerating a limited experiment in giving parliament a service that could be "used by the opposition". The political conditions that enabled these mid-80's initiatives to succeed in Europe underscore the importance of local and contingent elements; when it arose in Europe – unlike what occurred in the US in the Sixties –, it triggered a "hope" that technology would enable companies to become more competitive. Therefore what some have called the classic paradigm of TA (Van Eijndhoven 1997),

Table 1 Institutions and offices providing scientific advice and Technology Assessment (TA) to some European Parliaments

Country	Acronym	Name	English translation	Year of creation-operation
Denmark	DBT	Teknologi-rådet	The Danish Board of Technology	1986
European Union	STOA	Scientific and Technological Options Assessment at European Parliament	Scientific and Technological Options Assessment	1985–88
Flanders region	viWTA	Samenleving en Technologie	Flemish Institute for Science and Technology Assessment	2000–2
France	OPECST	Office Parlementaire d' Evalutation des Choix Scientifiques et Technologiques	Parliamentary Office for Evaluation of Scientific and Technological Options	1983–85
Germany	TAB	Büro für Technik-folgen-Abschätzung beim Deutschen Bundestag	Office of Technology Assessment at the German Parliament	1990
Switzerland	TA-SWISS	Centre d'évaluation des choix technologiques Zentrum für Technologie-folgen-Abschätzung	Centre for Technology Assessment (CTA) at the Swiss Science and Technology Council	1991
The Netherlands	Rathenau	Rathenau Instituut	Rathenau Institute (former NOTA)	1986
United Kingdom	POST	Parliamentary Office of Science and Technology	Parliamentary Office of Science and Technology	1986–89

which focused on the "early anticipation of adverse consequences", lost ground in Europe to other models such as the attempts to adapt the OTA paradigm. This paradigm focuses on clarifying options and the provision of S&T advice, basically geared to parliament (France, UK, Germany, European Parliament), or to models that emphasised the democratic control of technological developments and broader grass-roots involvement in decision-making (Netherlands, Denmark). Whatever the case, one of the key reasons behind the promotion of TA and POTAs in Europe is a hope that technology can further contribute to enhancing people's living conditions and boosting competitiveness, and perhaps that is why it was more closely connected to the policy process, albeit in a relatively marginal place such as parliament.

It must be emphasized that not all the initiatives to create parliamentary information and advice mechanisms were successful. In fact the difficulties involved in going beyond rhetoric and creating real and effective mechanisms, still persist in some of the POTAs that were equipped with scant resources, just as parliaments are poorly equipped in comparison to the executives. Certain initiatives failed, albeit on a temporary basis, one example being the case of Germany, whose first initiatives date back to the mid-Seventies, the proposal having been explicitly rejected in 1982. Others, such as the 1989 proposal to set up a POTA in the Spanish Parliament, (Quintanilla 1989) failed to get off the ground when early elections were called and the policy entrepreneurs changed. The causes, which are closely linked to parliamentary and political life, refer to the specific parliamentary sessions during which the initiatives were put forward; the fact that some MPs leave office whenever the political cycle changes means that sometimes these initiatives have failed, but also that they can be put forward again.

Attention should also be drawn to the contingency of POTAs, and of TA centres, in Europe. The same political dynamics that created the conditions for TA's institutionalisation may have created the conditions for its demise[4] (Bimber 1996). These cases go to show that if POTAs' opportunities of emerging hinged on political factors, such as the institutional reinforcement of parliament vis-à-vis the executive, and the will to make TA part and parcel of parliamentary life, their chances of survival are conditioned by the local adaptive strategies that they adopt.

4.2
The models in their political context

Despite the extraordinary variety of institutional arrangements that govern how POTAs are run, a functional analysis will show that there are two main models (Petermann 2000), in terms of their relationship with parliament, that reflect the key mission/s of each case: *instrumental* and *discursive*. The instrumental model (which could have applied to the defunct OTA) includes the POTAs whose chief (or only) customer are the respective parliaments or their committees. This is the case of POST, TAB, STOA, viWTA and OPECST; more recently, similar initiatives have been launched in countries such as Italy and Greece, but they are less active. The discursive model applies to the countries that have a long-standing civic tradition and have asked the TA institutes not only to contribute to "enlighten" parliamentarians' (and even the government's) decision-making processes, but also to help their respective societies to foster a social debate about the acceptability of technologies. This second type of organization, whose customers are not limited to parliaments, include the DBT, Ratheneau, and TA-SWISS. The three agencies' remits include the possibility of also advising their respective governments.

Broadly speaking, the political systems in which these european POTAs operate are characterized by a more formal than actual separation of powers, so the legislature and the executive tend to overlap, and in fact many ministers are also members of parliament. They are also characterized by a strong degree of party discipline in the parliamentary arena. Similarly, one finds a considerable degree of concentration of legislative initiatives within the government or the executive. Furthermore, parliamen-

---

[4] The issue is an important one, due to the government decision to cease funding the Baden-Württemberg Centre for Technology Assessment in 2003, or the pressures that the Danish Board of Technology has had to put up with since a conservative majority coalition took office, for the first time in many years.

tary committees usually mirror the distribution of power of the respective parliaments, and their role tends to be limited to preparing and drafting legislation, while the Plenary Session has the last word. All these characteristics portray a context marked by the centralization of power, so the differences in our cases in the frequency with which the POTAs inform either the parliamentary committees, or individual MPs or the Plenary Session, might not be very relevant from the impact viewpoint.

On the issue of differences, and the variables linking the political system to the POTAs' institutional arrangements, stable majority governments (either of one party or with very small coalitions) are more frequent in the United Kingdom and Germany, while minority and/or multiple coalition governments are more usual in Flanders, Denmark, Netherlands and Switzerland. It is also to be noted that interest groups' and stakeholders' relative capacity to influence the political system is very considerable in Flanders and Switzerland, strong in Netherlands, Denmark and Germany, yet more moderate in the United Kingdom. A political spectrum fragmented into several represented parties would seem to favour the emergence of discursive POTAs, insofar as one would expect the beneficiaries of the TA output to be widely distributed (and not concentrated) between the parties and society.

Some empirical evidence has been gathered in order to provide a certain amount of information about the TA production process[5]. The POTAs studied face a major dilemma. First of all, and as far as the scope of their activities is concerned, one might think that generalist organizations are at an advantage to specialist organizations, which would have to compete with regulatory agencies and others producers of very specific TA. Yet at the same time, in Europe these organizations have always had very limited budgetary and human resources, and when resources are scant, sometimes it is better to specialize. Most of the cases we have studied are generalist organizations, and all operate in social contexts in which TA development capacities are widely distributed, so no single organization enjoys a monopoly of this mission. However, several of the POTAs work exclusively or mainly for parliament (TAB, STOA, POST and viWTA), and the last three depend directly on it; whether or not this more restricted relationship results in a greater impact on some of the roles, is something to be verified empirically. Be that as it may, STOA and POST are supervised by their respective parliaments, while in all the other cases, the "administrative" supervision or dependence (for core funding) is linked to some ministry, normally the one responsible for science-related matters.

Inside the political system, POTAs must compete with other bodies that produce information about scientific and technical matters. These POTAs, which are generalist organizations, have to exist alongside more specialist organizations such as regulatory agencies or committees linked to the executive or the administration, and lobbies that produce information associated to certain interests. Despite certain differences existing in the size of POTAs, those we have studied have modest, limited resources and commonly have public funds as their main source of finance.

POTAs use scientific sources of information, although one should not underestimate the importance of non-scientific sources more related with the interest groups

---

[5] A survey was conducted among POTAs directors, using a semi-structure questionnaire covering a variety of topics about the mission, way of funding, resources and external links of their offices.

and civil society in general, especially in Netherlands, Denmark, Flanders and Switzerland, where their assigned missions are wider ranging. Even though it is the boards or councils of these organizations who decide what projects are taken on, there is a certain diversity in the degree of politicians' relative participation in these boards, which is greatest in STOA, in France, in POST and in viWTA. In all cases, the reports are widely distributed, although not of all these organizations have the promotion of social debate as part of their mission; it is in the case of Flanders, Denmark, Netherlands and Switzerland.

Finally, it must be stressed that none of these POTAs, regardless of the differences in the force of their political mandate, has been given anything resembling a power of veto, or even the compulsory role of informing certain legislation before it is passed. Their mission is limited to producing information and providing advice, and therefore in the absence of this type of institutionalised formal powers, the capacity to extend the bases of their internal and external ties, i.e., of networking, becomes very important, in particular with a view to their consolidation.

4.3
Adaptation strategies

Now we shall look at the incentives and opportunities that foster the interaction between politicians and parliamentarians on the one hand, and the POTA and its experts on the other. Some concepts of the organization theory apply to the analysis of POTAs which, once formed, face the challenge of surviving in local political environments, and therefore have to adapt their strategy and behaviour to their respective environment.

Very often, POTAs are initially approved for specific legislative periods, with reviews of their activities being scheduled beforehand, forcing these POTA to draw up adaptation strategies. Thus the first measure of a POTA's success is its capacity to develop strategies for adapting to the local political context, which enable it to survive and also to enhance its impact. We say adaptive success because they have survived reviews and also, in general, have turned into permanent parliamentary information production institutions.

Adaptation strategies are built upon the interaction between three factors: 1) the nature of the information required by politicians or legislators, or of the missions assigned; 2) experts' strategies in meeting these demands; 3) institutional arrangements. Thus there are two relevant issues: firstly, the attributes of the decision-making process and of the role that information production plays in it and, secondly, institutional constraints.

Adaptation strategies are based on identifying the incentive structures to which these POTAs must respond (the needs of the parliamentarians who act as their "principal") and keeping to the mandate established in its mission[6]. The adaptation strategy will depend on the power structure and its distribution: first of all, on the

---

[6]   Though relations can be explored on the basis of the principal-agent relationship idea, if one goes beyond the metaphor, the problem is that determining who the principal is in each case entails taking for granted that parliamentary structures are similar. Parliament is a principal, generally represented by the speaker, but there may be other principals within parliament, which is not a hierarchical body.

power/subordination relations between parliament and the executive; secondly, on the centralization of power within parliament, or on its decentralization; and lastly, on the type of autonomy/dependence (authority) relationship between the POTA and its parliamentary principal. Hence the reason for a TA's preferential connection with parliament can only be explained if one understands its nature as an information production process constrained by the political and parliamentary system's institutional arrangements.

In European parliamentary systems, it is governments who have the legislative initiative and, more often than not, parliamentary majorities serve to support government projects. Whatever the case, consideration must also be given to the parliamentarians' objectives, and their relationship with the parliament's functions in relation to the government. If the parliamentarians' essential objective is re-election, and this depends on their performance in their constituencies or with regard to their party, the tendency will be to legislate or to influence legislation; in european parliamentary systems, this means negotiating with the executive. Parliamentarians may also seek to enhance public policy, in which case the tendency to exercise the mission of controlling the government creates a context that is more favourable to the use of POTA-produced information[7].

One of the essential aspects of the institutional arrangements governing interaction between the demands for information and the POTA's capacity to supply it, is the extent to which power within a parliament is highly centralized or, on the contrary, more distributed; this is essential for understanding the nature and type of relations that may exist between parliaments and POTAs[8].

If a parliament's regulations vest the bureau and the house with all the powers, then the legislative committees have little authority and depend heavily on the political parties, which find it easier to control discipline in the house. For example, in the Spanish Parliament, the committees act "by delegation" from the house and, actually, tend to be the mechanism where the legislative proposals put forward by the Spanish Government are implemented if there is a large enough parliamentary majority. On the other end of the spectrum, in a European Parliament-type structure, it is essentially the committees that "prefigure" the house's decision, because they prepare and hold lengthy discussions about the different matters on the agenda, and party discipline is very limited.

When parliaments and their committees make decisions, or when parliamentarians form their opinions, these are generally based: on their political preferences; public opinion; the stance taken by the parties, especially those that support the government and above all on the information produced by the bureaucrats who serve the executive; and perhaps on the information produced by interest groups.

Their success and impact depend on how they adapt to the environment; but the POTA's model of organization will mirror the model of "centralization of power". If power is highly centralized (as tends to occur in European Parliaments), there are

---

[7]   It would be interesting to ascertain to what extent the type of specific information products provided by the POTAs makes it easier for parliamentarians to "legislate" or to "control"; in principle, given that TA is a slow process, we are inclined to think that it is more relevant in the control tasks.

[8]   The number of different political parties could provide an alternative explanation for the decentralization of power in parliament. Germany (4/5), Denmark (8), Holland (9/10).

fewer chances of consolidating a pluralist clientele. The more power is centralized within parliament, the more demand for "policy advice" is likely to become politicised, and the more likely it is that information will give rise to back-up information rather than debate. The provision of information will tend to come from the executive, supported by the parliamentary majority.

Events at the OTA have shown that, of the two models described above, the instrumental model has the greatest impact in a relatively decentralised power structure marked by division. In different circumstances, and given that none of these organizations have been given institutional powers to veto policies, this model may come up against political hurdles when it attempts to really influence decisions. Even though it cannot be denied that having Parliament as one's only or chief customer, in addition to a more specific mission, can be advantageous for gaining access at the right time.

The third factor associated with the structure of power and its distribution refers to a POTA's degree of autonomy or dependency with respect to parliament. Despite their acknowledged autonomy, the instrumental POTAs' dependence on the "authority" of the legislators produces a sense of "owner identity" that in fact constrains their autonomy, and can manifest itself in the selection of objects for analysis. If this argument is taken to the extreme, sometimes it is the majority parliamentary groups who decide how an instrumental POTA operates. In this context, following adaptation strategies devised to strengthen POTA experts' autonomy and independence is limited by the distrust that this can trigger among the principals. They will only support the POTA insofar as, apart from being neutral, non partisan etc., it is instrumental to their political goals. Remaining loyal to the principal fosters trust and that, in theory, is the decisive factor for increasing access.

However, the fact is that politicians inevitably have countless alternative sources of information and decision-making criteria. Besides, having only one principal, as occurs with instrumental POTAs, makes them extraordinarily dependent upon their functionality.

In the case of discursive POTAs, the parliamentary information and advice service is provided by organizations that are significantly more independent and autonomous. Having a large number of missions, i.e., principals (parliament, the public at large, sometimes the government), makes them extraordinarily more independent. The slip side is that, even when there is a minimum degree of neutrality, non-partisanship, etc., there is unquestionably less of a sense of "owner identity" regarding the TA results. Furthermore, these institutions' capacity allows them to manoeuvre socially in order to further the proposed objectives directly.

Theory has it that the less centralized a parliament's power structure is, the more likely a POTA will be to establish itself and have an impact. By impact we mean the ability to have a bearing on socio-political change, or on the change in policies. So far we have seen how factors conditioning the distribution of power essentially constrain POTAs' opportunities to access and have an influence on parliamentary decisions, and now we must look at a second issue, namely POTAs' capacity to manoeuvre with respect to the world of social and political actors outside of parliament. In institutional contexts where political divides and decentralization are not so common, a POTA's impact largely hinges on its ability to build coalitions that defend and support its information products. This entails mobilizing actors outside parliament, which is difficult and can trigger institutional conflicts with a POTA's principal, namely the parliament.

A large body of literature has taken the "advocacy coalition" (Sabatier 1988) perspective as the theoretical framework for analysing public policies, and especially their changes. Some of the theory's premises can be very useful for the subsequent empirical analysis of how the TA organizations' impact differs in line with the dominant institutional models that we have outlined. First of all, this perspective focuses on the relevance of S&T information in the political process and in the change of certain policies. Secondly, any impact on changes in policies can only be broached from a broad time perspective of several years. In other words, applied to our case, coalition-building is associated to the prior institutional consolidation of the TA organizations. There are several reasons for this, but perhaps the most significant is that these organizations must be capable of gaining recognition and visibility before they set up networks. Thirdly, any analysis conducted to ascertain why policies change in our societies should not be limited to a single organization, be it governmental or of any other type, but to the subsystem of that policy, or what has become known as "policy domain". Besides, the concept of "policy domain" must be broadened to include not only the traditional triangle formed by the administration, legislature and interest groups, but also two other categories of actors: on the one hand, the media, academia, and political analysts, and on the other, the subnational and supranational levels of government, which sometimes give rise to innovations and disseminate ideas.

Another question is the stability of the effects and their duration. Institutional arrangements permit different degrees of legislative change: much less so in a system with a strong separation of powers than in a parliamentary system. A POTA's TA product or process may influence political options, but the effects not necessarily remain in time. In parliamentary systems in which a majority holds most power, the majority coalition or party can change not only policies but even legislation many times, provided that they consider that this will not overly damage their electoral gains in the medium term. This is where the breadth of coalitions comes into play and where the apparent advantage of enjoying direct access to parliamentarians and the sense of owner identity regarding its products can become an obstacle. TA stems from and is institutionalised on the basis of relations between technology and society that are always complex and sometimes opaque; the organizations that produce TA will have an impact as long as there are enough actors who perceive their effects as positive, not only because they clarify options, but because they legitimate some of them through their evaluation. TA can create or change the point of view not only of politicians, but also of industry or of the public at large. External ties matter because the power bases that are behind the several options are very often external to the political system.

In this section we have shown that is possible to predict the evolution of POTAs or at least, given the institutional context, their best adaptive strategy. Straying from that optimum strategy can trigger conditions that jeopardise their very survival. Furthermore, this analysis has provided the key elements of an institutional theory regarding the factors that determine how TA activities are likely to impact the political system and society.

4.4
The impact of TA in parliament

One might suppose that the impact of TA activities is the direct result of the "quality of information", of the "professionalism of the producers of that information", of the unique nature of the subject matter, or of any other intentional or simply voluntarist variable. Our argument, however, is that *the institutional context* in which TA takes place, the *modes of organization of TA production*, and the interaction between politicians' demands and the POTAs' experts' strategies, are key factors in the general explanation of the levels of impact and the varieties thereof. Then we argue that the institutional arrangements that govern POTAs, the rules of the political game and the incentive structure that POTAs face, are key factors for explaining the depth and extension of the impact of their TA activities[9], as well as their adaptation and consolidation. In this respect, we would emphasize the contradictions that sometimes exist between the institutional arrangements and the relative influence of the technological assessment of the organizations linked to the parliaments. The two institutional properties discussed in the previous section: autonomy/dependence and exclusivity/non-exclusivity of the assessment for the parliaments are essential for understanding the diversity of impacts.

We aim to construct an analysis that lets us associate the probability of a given type of impact (see the typology) to the dominant types of institutional arrangements (and the access and coalition structures that they permit or foster); however, we might also understand that institutional structures are "prerequisites" but do not suffice to predict the type of impact and the results in each specific case. This exercise does not aim to explain each and every one of the cases of TA reports and their impact, but basically to use empirical analysis to put forward a normative proposal that lets us understand the expected impact of TA activities, and at the same time that lets us reinforce or maximize the influence of its products.

In the previous section we have shown that POTAs' impact is the result of the combination of two main variables: the first is the organizations' capacity to build broad support coalitions both inside and outside the legislature, and the second is their capacity to access decision-makers in a context of competition with other organizations that also produce evaluative information about technological options. Instrumental and discursive POTAs have both types of these capacities, in differing degrees, and this degree depends in turn on the institutional context, the rules of the game and the incentive structure within which these organizations operate.

It seems clear that the POTAs' degree of autonomy-dependence means that, in principle, the instrumental types enjoy more direct access to parliamentarians; yet at the same time, to avoid jeopardising their survival through disloyalties to their principals, less capacity to engage in building coalitions "outside" parliament. On the flip side, the more discursive POTAs' direct access to parliament is more conditioned by the lack of a sense of "owner identity", because these institutions can

---

[9]   The explanation is essentially probabilistic, because in specific circumstances, other general explanatory variables (political leadership, changes of political juncture, etc.) or regarding the nature of the specific policy domain affected by the TA, may also serve to interpret the situations.

work for other principals; in compensation, their greater autonomy enables them to foster the mobilization of support coalitions.

If empirically the two aforementioned dimensions occurred at the same time in a certain type of TA organization, predicting its relative impact would be very simple and linear. However, this would only be imaginable in a world where politicians only cared about the truth and about improving public policies. In the real context of political systems and, in particular, of parliamentary ones, with the actual incentives and restrictions, a contradictory relationship exists between those two dimensions; in other words, if a greater capacity to build broad coalitions were associated inversely to the capacity to gain more direct and restricted access to legislators, a clear relationship could not be said to exist between a specific POTA model and a greater impact. What we argue is that not only can those contradictions occur, but also that the relationship is far more complex, and controlled by variables associated to the specific institutional arrangements, such as: political centralization or decentralization in the legislature, the relative power of the parliamentary committees, the strength of party discipline, the power of the political mandate given to these organizations when they are created, and by the dynamics by which the politicians obtain legitimacy for their decisions. In addition to all these variables that are external to TA organizations, there are other internal factors, related to these organizations' information production process, which have to do with organizational capacities (budgetary and human resources) that must also be borne in mind when analysing the impact.

The coalitions' breadth and access matter because the impact of TA, or any policy advice, is also conditioned by the information communication structure. That structure is what connects those who produce the information and those who use it or its customers. Be that as it may, what seems certain is that, in absolute terms, the broader the "coalition" of political, social and media actors, the bigger the impact of the TA (science-based knowledge for political use in non-academic communication formats) produced by specialised institutions. The greater the inclusivity in the number of actors (stakeholders) associated to the "technology" or to the "problem" who assimilate the proposal, the greater the impact[10].

Impacts can be classified according to the typology[11] we present in table 2. We will now relate this with the institutional types that we have developed, in order to at least generate a few descriptive or normative hypotheses as to how POTAs can augment their impact. The implicit assumption of this typology of impacts is that they are not constrained to the "change in policies", but instead can be limited, in a gradient, to "raising knowledge" or "framing the problem". It might seem that the only "justification" of POTAs is that they produce substantive third dimension impacts, yet depending on the dominant institutional arrangements, it may suffice if, with the "information's owner identity" it achieves impacts associated to raising knowledge or forming of opinions.

---

[10] Some case studies based-analysis has found that the strength of the OTA analytic process was its emphasis on broad participation. This practice increased the likelihood of impact by involving many of the actors crucial for an effective "distribution network" (Whiteman, 1997, p. 188).

[11] This typology has been developed by the research team of the TAMI project.

Table 2 Typology of Impacts

Impact Dimension / Issue Dimension	I. Raising Knowledge	II. Forming Attitudes/Opinions	III. Initialising Actions
Technological/ Scientific Aspects	Scientific Assessment a) Technical options assessed and made visible b) Comprehensive overview on consequences given	Agenda Setting g) Setting the agenda in the political debate l) Stimulating public debate n) Introducing visions or scenarios	New R&D Policies v) New action plan or initiative to further scrutinize the problem decided q) New orientation in policies established
Societal Aspects	Social Mapping d) Structure of conflicts made transparent	Mediation h) Blockade running i) Bridge building j) Self-reflecting among actors	New Decision making Processes u) New ways of governance introduced w) Initiative to intensify public debate taken
Policy Aspects	Policy Analysis e) Policy objectives explored f) Existing policies assessed	Re-Structuring The Policy Debate k) Comprehensiveness in policies increased p) Policies evaluated through debate o) Democratic legitimisation perceived	New Policies r) New legislation is passed s) Policy alternatives filtered t) Innovations implemented

According to this typology of impacts, POTAs can be said to have succeeded in the traditional mission of providing information for the political process. The review of their work point to a bigger impact on the roles included in the "knowledge production" dimension than on the other two dimensions. However, it must be added that a POTA's organizational age is a good predictor of a growing impact on this traditional role.

Conversely, we have found less of an impact on the roles included in the third dimension, "initialising actions", than in the other two. Therefore, in accordance with one of our initial hypotheses, the more constrained ties between certain POTAs and their respective parliaments do not seem to guarantee a bigger impact on legislation or on the change of policies in contexts marked by the centralization of power and control by majority. Perhaps the fact that the structure of Europe's parliaments is more prone to promoting the function of "controlling" a government's action rather than legislation, explains the limited impacts, on the basis of this institutional constraint. There is no question that the lack of a strong political mandate, with a certain veto capacity, has something that do with this result. Yet this is only part of the explanation.

Furthermore, the three dimensions of the impacts must be considered from a time perspective. In time, the organizations studied seem to have consolidated the roles related to the production of knowledge, as demonstrated by the hundreds of briefings, notes, and dozens of reports, hearings, public debates or consensus conferences. Nonetheless, the impact on the first dimension roles are closely linked to capacities and resources, in particular scientific advice and policy analysis in specific matters that require a high level of internal or external expertise[12]. The quality of these products depends heavily on available funding, but their impact also depends on the relative use that decision-makers make of alternative sources of information production. However, with equal resources, discursive-type POTAs may have a bigger impact on the knowledge production role known as "social mapping" (the role whose output is to outline the structure of conflicts) due to factors basically related with the process (method) by which the information is produced. The impact in this dimension depends on the sense of owner identity that politicians and parliamentarians afford this information, on their trust in the POTA, which is divided between institutional neutrality and loyalty to the parliaments.

The importance of the time perspective is evident due to the simple fact that information production is a precondition for influencing the opinions and attitudes of the actors involved. Influencing the agenda is one of the roles within this second dimension, and there are several ways of doing this, but one traditional one is "framing the problem"(Schön and Rein 1994) or structuring the issue. Influencing the agenda means that the decision-makers accept that the information must be considered politically. Gaining access to decision-makers and attracting their attention can be a determining factor, especially when different sources of information are vying to attract politicians and legislators' attention. Once again, the POTAs that respond to a model in which external ties play an important role in the information production process, such as discursive POTAs, are likely to have a bigger impact on other roles of this dimension. These include brokering and communication between actors (breaking deadlocks in dialogue, bridge-building) or stimulating of a debate in which the options are assessed, because they have access to actors other than politicians.

The instances of strong political impact, which pertain to the "Initialising actions" dimension, are few and far between. Mention has been made of a DBT

---

[12] According to some role descriptions made by TAMI, in certain areas it can be difficult to find competent, independent experts: e.g.: FX nuclear power, pharmacy).

report on "food and genes", which sparked a political debate that led to legislation on genetically modified food. However the DBT interviewees acknowledge that it is very difficult to attribute the legislation to the TA's impact, i.e., to isolate that cause. Other examples of outputs reviewed on "policy alternatives filtered", are also Danish. The limited examples in this dimension seem to indicate that having a second dimension impact, i.e. the change in attitudes and opinions, can be a precondition for having an impact on policies and legislation.

TA organizations seem to have established themselves, and have survived minor changes and "reviews" of their activities, in which they have received support. The DBT "crisis" or the recent demise of the Baden-Württemberg CTA point to the unstable balances that TA organizations face in the context of the political system.

4.5
Conclusions

Having reached this point, it is time to draw a few normative conclusions. In the context of today's political systems, the two types of POTA face such a dilemma about their future that one could say that they are not balanced systems, even if they have attained a certain degree of recognition and institutionalisation. The advantage of the more instrumental varieties, which tend to be subject to greater political authority, is that they have direct access to decision-makers. But the way in which the latter demand the information only contributes decisively to their impact when the information has a clear "owner identity", as compared to external, non-parliamentary sources. Developing an owner identity entails building up parliamentarians' trust (both political and technical) in the POTAs. This is achieved through loyalty, i.e., by POTAs limiting their own capacity for initiative, especially if this is not consistent with the current parliamentary majorities. POTAs that are overly active in promoting TA options run the risk of jeopardising the institution's sense of collective identity and trust; if the activism is or has been consistent with the parliamentary majority, then the risk might occur if there is a change of majority that may have regarded the POTA's activism as too instrumental in favour of the government majority. By way of example, POST and TAB grew stronger during "left-wing" majorities. So forcing the impact, especially in the initiatives dimension, can put the organization's very survival at risk in the medium term.

Furthermore, the more discursive POTAs, which are less instrumentalised by their respective parliaments, have greater scope for manoeuvring to enhance the impact of their TA output, because access to social actors is one of their institutionalised missions. They seem to have less of a direct influence on parliamentarians, in information terms, insofar as the sense of ownership of the information is not so intense as in the instrumental POTAs. Although their influence is potentially larger due to their capacity to build social coalitions in the public opinion that change the political vision in a specific field. The risk that these POTAs may face is that they are nothing more than "independent institutes" that depend on the national budget and may end up paying the price of their autonomy and independence if their activism upsets the political preferences or interests of the politicians in office at a given moment. The development of discursive TA approaches in civil society, supported by private establishments, might be a justification for stopping their public support.

Table 3 Descriptive and normative hypotheses regarding the survival and impact of POTAs

- Quality, neutrality and external communication are factors that influence the impact on all the aspects considered.
- Quality depends heavily on organizational capabilities (budgetary, human, etc), and neutrality depends on access to sources of independent expertise.
- The scope of communication depends on the density of the external networks (with academia, the Media, stakeholders, industry) that allows the autonomy to be increased.
- The impact on some roles is closely related to the participatory information production processes (methods), which call for a certain density in the external networks.
- A specific mandate and close ties with the institutions make it easier for POTAs to gain access to politicians, due a sense of owner identity, as opposed to competitor organizations
- The instrumental type POTA model is related with a more specific mandate and a potentially more direct access
- Capturing politicians' attention depends not only on the access but also on the extent to which the former see gains in terms of legitimacy for their decisions
- The capacity of paying attention to or receiving information is conditioned by trust in the source, and this trust increases with the sense of identity and "exclusiveness"
- The legitimacy of decisions is related to the scientific-technical quality of the information input and also with the breadth of the social base behind them (transparency).
- In a political system marked by the centralization of power in executive, with Parliaments dominated by majorities or coalition governments, with little separation of powers and where committees mirror the distribution of power in the house, the POTAs stand the best chances of survival and impact if they form support coalitions that include outside actors.
- The discursive model of POTA, with more general missions that imply more external relations, is more likely to be able to form such coalitions.
- The risk of transforming either of the two models into activism is that it is subject to the costs and benefits of a political alignment subject to changes.

If these organizations focused on low profile missions (as a documentation service), perhaps they would run less of a risk, but their impact might be marginal, because it entails complete submission to political authority. If the information functions multiply, then maintaining it calls for technical discipline (neutrality, etc.) and an institutional identity that can only be maintained through less autonomy and loyalty to the institution.

If the option is to "act", by building coalitions to enhance the impact, then there is no risk in the short term if they align with the "parliamentary majority"; in the medium term, however, a new majority can make one "pay" for institutional disloyalties. In short, the shift to activism entails remaining subject to the costs and benefits of the political alignment.

The fact is that some of the best adaptation strategies that POTAs use to improve their chances of survival clash structurally with the desire to increase the direct impact of their TA activities on policy-making activities.

Lastly, it is important to understand how the organization of power (the extent to which it is centralized) conditions the POTAs' adaptation strategy. The more concentrated power is, the higher the "risk of partisan policy or politicisation", and the more likely they are to survive in the short term. At the same time, the more power that the (few and disciplined) parties and the Speaker of the House have than the committees, the fewer opportunities they will have to build up broad supporting customer bases inside the parliament (see table 3).

References

Barker A, Guy Peters B (eds) (1993) The Politics of Expert Advice: Creating, Using and Manipulating Scientific Knowledge for Public Policy. University of Pittsburgh Press, Pittsburgh

Berry FS, Berry WD (1990) State Lottery adoptions as policy innovations: An event History analysis. The American Political Science Review, vol 84,2, jun, 395–415

Bimber B (1996) The Politics of Expertise in Congress. The Rise and Fall of the Office of Technology Assessment. State University of New York Press, Albany

Bimber B, Guston DH (1995) "Politics by the Same Means. Government and Science in The United States". In: Jasanoff S, Markle GE, Petersen JC, Pinch T (eds) (1995) Handbook of science and technology studies. Sage, London, 554–571

Bimber B, Guston DH (1997) Introduction: The End of OTA and the Future of Technology Assessment, Technological Forecasting and Social Change, vol. 54, issues 2–3, February–March 1997, 125–130

Bromley DA (1995) The President's Scientists. Yale University Press, New Haven

Coleman J (1966) Medical Innovations: A diffusion Study. Bobbs-Merrill, New Yorks

Coleman J, Katz E, Menzel H (1957) The Diffusion of an Innovation Among Physicians. Sociometry, Vol. 20, No. 4. (Dec. 1957), pp 253–270

Collier D (1993) "The Comparative Method". In: Finifter AW (ed) (1993) Political Science: The State of the Discipline II. APSA, Washington DC

Cooper J, Brady DW (1981) Towards a diachronic Analysis of Congress. The American Political Science Review, vol 75, 4, Dec, 988–1006

Cozzens SE, Woodhouse EJ (1995) "Science, Government, and the Politics of Knowledge". In: Jasanoff S, Markle GE, Petersen JC, Pinch T (eds) (1995) Handbook of science and technology studies. Sage, London, 533–553

Emerson RM (1962) Power-dependence Relations. American Sociological Review, vol 27, 1, Feb, 31–41

Feldman MS, March JG (1981) Information in Organizations and Signal and Symbol. Administrative Science Quarterly, vol. 26, pp 171–186

Finnemore M (1996) National Interest and International Society. Cornel Unievrsity Press, Ithaca (NY)

Gibbons JH, Gwin HL (1988) "Technology and Governance: The Development of the Office of Technology Assessment". In: Kraft ME, Vig NJ (eds) (1988) Technology and Politics. Duke University Press, Durham, 98–120

Haas PM (1992) Introduction: epistemic communities and international policy coordination. En International Organization vol 46, n.1, pp 1–35

Heclo H (1974) Modern Social Politics in Britain and Sweden. Yale University Press, New Haven (Co)

Hedström P (1998) "Rational imitation". In: Hedström P, Swedberg R (eds) Social mechanism. An Analytical Approach to Social Theory, Cambridge University Press, Cambridge, pp 306–327

Holdsworth D (2000) "Parliamentary Technology Assessment by STOA at the European Parliament". In: Vig N, Paschen NH (eds) (2000) Parliaments and Technology. The Development of Technology Assessment in Europe. State University of New York Press, Albany (NY), 199–226

Holt RT (1977) Technology Assessment and Technology Inducement Mechanism. American Journal of Political Science vol 21, 2, May, 1977, 283–301

Jasanoff S (1990) The Fifth Branch: Science Advisers As Policymakers. Harvard University Press, Cambridge

Jasanoff S (1995) Science at the Bar: Law, Science and Technology in America, Harvard University Press, Cambridge (Ma)

Jones M, Guston DH, Branscomb LM (1996) Informed Legislatures. Coping with Science in a Democracy. University Press of America, Boston

Killian JR (1977) Sputnik, Scientist and Eisenhower. The MIT Press, Cambridge (Ma)

Kingdon J (1984/1995), Agendas, Alternatives and Public Policies, Little Brown, Boston

Klüver L (2000) "The Danish Board of Technology". In: Vig N, Paschen JH (eds) (2000) Parliaments and Technology. The Development of Technology Assessment in Europe. State University of New York Press, Albany (NY), 173–197

Kuhlman S, Boekholt P, Georghiou L, Guy K, Héraud J-A, Laredo P, Lemola T, Loveridge D, Luukkonen T, Polt W, Rip A, Sanz-Menéndez L, Smits R (1999) Improving Distributed Intelligence in Complex Innovation Systems, Final report of the Advanced Science & Technology Policy Planning Network (ASTPP), Brussels, EC-DGXII, mimeo

Laurent M (2000) "France: Office Parlementaire d'Evaluation des Choix Scientifiques et technologiques". In: Vig N, Norman J, Paschen H (eds) (2000) Parliaments and Technology. The Development of Technology Assessment in Europe. State University of New York Press, Albany (NY), 125–146

Majone G (1991) Cross-National Sources of Regulatory Policymaking in Europe and the United States. Journal of Public Policy 11, 1, January–March 1991, 79–106

Meyer JW, Hannan MT (1979) National Development and the World System. Education, Economic and Political Change 1950–1970. The University of Chicago Press, Chicago

Meyer JW, O Ramirez F, Soysal YN(1992) World Expansion of Mass Education, 1870–1980. Sociology of Education vol 65, 2, April, 128–149

Norton M (2000) "Origins and Functions of the UK Parliamentary Office of Science and Technology". In: Vig N, Paschen JH (eds) (2000) Parliaments and Technology. The Development of Technology Assessment in Europe. State University of New York Press, Albany (NY), 65–92

OTA-Office of Technology Assessment. United States Congress (1996) OTA legacy. US Printing Office, Washington DC, 5 vols (Cdrom)

Paschen H (2000) "The Technology Assessment Bureau of the German Parliament". In: Vig N, Paschen JH (eds) (2000) Parliaments and Technology. The Development of Technology Assessment in Europe. State University of New York Press, Albany (NY), 93–124

Petermann T (2000) "Technology Assessment Units in the European Parliamentary Systems". In: Vig N, Paschen JH (eds) (2000) Parliaments and Technology. The Development of Technology Assessment in Europe. State University of New York Press, Albany (NY), 37–61

Polsby NW (1968) The Institutionalization of the U.S. House of Representatives. The American Political Science Review vol 62, 1, March, 144–168

Polsby NW (1984) Political Innovation in America. The Politics of Policy Initiation. Yale University Press, New Haven(Co) – London

Polsby NW, Schickler E (2002) Landmarks in the Study of Congress since 1945. Annual Review of Political Science vol 5, 333–367

Porter AL, Rossini FA, Carpenter SR, Roper AT (1980) A Guidebook for Technology Assessment and Impact Analysis. North-Holland, New York

Przeworski A, Alvarez ME, Cheibub JA, Limongi F (2000) Democracy and Development. Political institutions and well-being in the World (1950–1990). Cambridge University Press, Cambridge

Quintanilla MÁ (coord.) (1989) Evaluación Parlamentaria de las Opciones Científicas y Tecnológicas. Centro de Estudios Constitucionales, Madrid

Rip A, Misa TJ, Schot J (eds) (1996) Managing Technology in Society: The Approach of Constructive Technology Assessment. Pinter, London

Rogers EM (1962/1995) Diffusion of innovations. The Free Press (4ª ed.), New York

Rose R, Davies PL (1994) Inheritance in public policy. Yale University Press, New Haven

Rose R (1991) What is Lesson-Drawing?. Journal of Public Policy 11, 1, January–March 1991, 3–30

Sabatier P (1988) An Advocacy Coalition Framework of Policy Change and the Role of Policy-Oriented Learning Therein. Policy Sciences vol 21, 129–168

Schneider M, Teske P, Mintrom M (1995) Public Entrepreneurs. Princeton University Press, Princeton

Schön DA, Rein M (1994) Frame reflection. Towards the resolution of intractable policy controversies. Basic Books, New York

Simon H (1983) Reason in Human Affairs. Stanford University Press, Stanford

Smith BLR (1992) The Advisors: Scientist in the Policy Process. Brookings Institution, Washington D.C.

Smits R (1987) "Aspects of the integration of science and technology in the American society". In: VV.AA. (1987) Technology Assessment. An Opportunity for Europe. The Hague: Government Printing Office

Smits R, Leyten J, Den Hertog P (1995) Technology assessment and technology policy in Europe: New concepts, new goals, new infrastructures. Policy Sciences, 28, 3, August, 271–299

Sundquist J (1981) The Decline and resurgence of Congress. The Brooking Institution, Washington DC

The IPTS Report (2001) Special Issue: The Provision and Impact of Scientific Advice. The IPTS Report n° 60, December 2001, Sevilla: CE-CCI_IPTS

The IPTS Report (2003) Special Issue: Assessing the Impact of Scientific Advice. The IPTS Report n° 72, March 2003, Sevilla: CE-CCI_IPTS

Van Eijndhoven J (1997) Technology Assessment: Product or Process? Technological Forecasting and Social Change vol. 54, issues 2–3, February–March 1997, 269–286

Van Eijndhoven J (2000) "The Netherlands: Technology Assessment from Academically Oriented Analyses to Support of Public debate". In: Vig NJ, Paschen H (eds) (2000) Parliaments and Technology. The Development of Technology Assessment in Europe. State University of New York Press, Albany (NY), 147–172

Vig NJ, Paschen H (2000) "Introduction. Technology Assessment in Comparative Perspective". In: Vig NJ, Paschen H (eds) (2000) Parliaments and Technology. The Development of Technology Assessment in Europe. State University of New York Press, Albany (NY), 3–35

Vig NJ (2000) "Conclusions. The European Parliamentary Technology Assessment Experience". In: Vig NJ, Paschen H (eds) (2000) Parliaments and Technology. The Development of Technology Assessment in Europe. State University of New York Press, Albany (NY), 365–394

VV.AA (2000) EUROPTA. European Participatory Technology Assessment. Participatory Methods in technology Assessment and Technology Decisión-Making. The Danish Board of Technology, Copenhagen

Walker JL (1969) The diffusion of innovations among the American States. The American Political Science Review vol 63, 3, Sept., 880–899

Whiteman D (1997) Congress and Policy Análisis. A Context for Assessing the Use of OTA projects. Technological Forecasting and Social Change vol 54, 1997, 177–189

Wolman H (1992) Understanding Cross National Policy Transfers: The Case of Britain and the US. Governance, An International Journal of Policy and Administration, vol. 5, n. 2, pp. 27–45

Organised Interests in the European Union's Science and Technology Policy – The Influence of Lobbying Activities

Theo Karapiperis[13] and Miltos Ladikas

1
Introduction

Most modern industrialised societies are governed by one type or another of *representative democracy*, in which the ultimate political power rests with *regularly elected and accountable institutions*. Elections are based on free party competition and the elected members of these institutions (governments, parliaments and other elected bodies) are supposed to represent the electorate in a balanced way that guarantees the peaceful resolution of conflicts to maximum overall benefit, and thus ensures the durability of the political system.

Representative democracy has reached such a level of maturity and self-confidence in modern industrialised societies, that it does not feel threatened, and even finds itself revitalised by the emergence of an important novel dimension to the already very complex organisational structure that underlies it. What has come to be known as *'civil society'* is no longer content with the privilege of regularly electing representatives that will defend its often divergent interests in local, regional, national or supranational bodies. It tries to intervene in the political process on a continuous basis via a multitude of private and public organisations representing the interests of different societal groups against those of state institutions, but also against each other.

Organised interests try to influence the political process in their favour by means of a variety of methods, including what has come to be known as *'lobbying'*. The word stems from 'lobby', defined as "a hall in a legislative building used for meetings between the legislators and members of the public" [14]. Lobbying is then defined as the "attempt to influence (legislators etc.) in the formulation of policy" [15].

The preferred field of lobbying activity is the *legislative process*, due to the more or less *binding* and *lasting* character of its outcome. However, modern society is in practice so complex that, even when it is governed by the *rule of law* at its best, the legislation adopted by its representatives cannot determine unambiguously and for ever the relative positions of all parties involved in cases of conflicting interests. There is therefore a wide scope for lobbying activity beyond the narrow field of legislation. Thus, much lobbying activity has a *pro-active* character and deals with obtaining and providing information, building and maintaining supporting relations and networks, ensuring the appointment of favourable individuals in important

---

[13] The views expressed are those of the author and do not necessarily reflect those of the European Parliament.
[14] Collins English Dictionary, Millennium edition, HarperCollins, 1998.
[15] ibid.

positions, influencing the process of agenda-setting, building alliances with key players, accelerating or braking certain developments, pursuing favourable administrative decisions (financial favours, awarding of contracts, ensuring desired outcomes in inspections or evaluations etc.).

Ensuring a desired outcome in a legislative or any other decision-making process is a extremely complex job. It calls for a very high degree of concentration and a great deal of ingenuity and imagination. One has to identify the right people in the decision-making process, understand their institutional role and behavioural patterns, approach them in an appropriate way, 'push' them towards the right direction, ensure the support of secondary actors and minimise the resistance of those that cannot be persuaded. All this happens in a dynamic context of shifting conflicts and alliances, in which, when an all-out *victory* is not possible, one may have to opt for the next best outcome (i.e. some kind of *compromise*) before it is too late.

What underlies an effective execution of such a complex task is *knowledge*. One has to know sufficiently well the political and institutional context, understand its dynamics and the role of different actors and master the necessary skills in psychology and communication (Bouwen 2002). All this is necessary, but worthless, if one does not also have the requisite *factual knowledge*. In the context of science and technology (S&T)-related issues, the knowledge needed is that possessed by specialised scientists and engineers. Given the remoteness of S&T-related facts from the standard knowledge base of most laymen, lobbyists have an opportunity of playing a useful role, but, for exactly the same reason, they also have the power to play an extremely destructive one.

A lobbyist engaged in a campaign on a S&T-related issue may be a specialist on the subject of interest and, if he/she feels that reality is on his/her side, can enrich the political decision-making process by supplying much-needed specialist knowledge to the actors involved in that process. It is common knowledge that most political decisions on specialised issues are taken by people who are laymen in the field of interest. It is therefore not surprising that a recent survey of the relations of Members of the European Parliament (MEPs) with interest groups found that MEPs value "gaining expert knowledge" as the primary purpose for meeting a lobbyist (Kohler-Koch 1997). This makes politicians amenable to very strong outside influences, as they often share with a large section of their electorate a sense of intellectual powerlessness with respect to the particular issues. As a result, they are often insufficiently equipped to distinguish an honest attempt to influence their position on a certain issue on the basis of genuine facts from a malicious manipulation based on a biased, distorted or completely false representation of those facts, motivated by the wish to defend certain organised interests at any cost.

The risk of being badly misled by fraudulent lobbyists leads many politicians to seek the support of more *independent* and therefore, more *trustworthy* experts. In many cases politicians build their own networks of experts, so that they can appeal to them in a flexible and efficient way whenever needed – European politicians do not usually possess the financial resources of US congressmen who can afford to employ directly high-level experts among their staff. This leaves sufficient room for a positive contribution from various *TA bodies*, whether independent or associated with specific institutions, such as parliaments. Such bodies have strong advantages with respect to specialised institutions or individuals with none or little TA expertise: They are familiar with the *political stakes*, have collectively a good overview of *related S&T issues* and generally provide better assurances of *independence* and *professionalism* in the *comprehensive* evaluation of S&T *options*. As far as TA bodies attached to particular parliamentary institutions are concerned, they constitute indispensable instruments in the task of safeguarding the collective decision-making *autonomy* of those institutions.

This paper aims at discussing the issue of the effects of lobbying activities in policy-making in relation to TA functions. The focus is at the European level of decision-making which is the most relevant and influential nowadays. The development of the discussion is based on empirical data deriving from the analysis of a survey submitted to members of the European Parliament and a number of case studies debated at the European Parliament.

2
The extend of lobby influence: the MEP survey

There is very little actual empirical research on the influence of organised interest in decision making at the European level. Theoretical work on organised interest in the EU (Kohler-Koch 1997; Hix 1999; Grande 2001; Van Schedelen 2002) is rarely accompanied by raw data from e.g. survey research. Presumably as the result of difficulties in reaching the target audience and the sensitivity of the matter, there is still lack of comprehensive data.

Perhaps the best data resource on the issue is a recent survey study contacted by the European Parliament Research Group (EPRG) on voting behaviour of MEPs, which also included many questions on the influence of organised interests in decision-making. The questions unveiled frequency of contacts with interest groups and influence of these groups in voting behaviour. The breadth of the sample (including all MEPs from the 15 EU Member States) and the response rate (more than 30% of MEPs) make this survey an interesting tool for comparative data analysis in the issue. Following are the results relating to the issue of organised interests[16].

On the question "How frequently are you in contact with the following groups, people or institutions", providing a list of alternatives including "organised groups" and "lobbyists":

- 57% of MEPs meet organised group representatives every week, while 91% meet them at least once a month;
- 44% of MEPs meet lobbyists every week, while 72% meet them at least once a month;
- The organised associations represented in frequent meetings (at least once a month) are: Environmental associations (51%), Professional associations (46%), Industry organisations (43%), Consumer associations (42%), Trade unions (39%), Trade and commerce associations (32%).

---

[16] Survey data kindly provided to the authors by Dr Simon Hix, Government Department, London School of Economics and Political Science, UK.

Furthermore, the survey question: "How often do you receive recommendations on which way to vote from the following parties or groups?" also included the response alternative of "interest groups":

– Voting instructions are received very frequently by the following groups: Interest groups (72%), EP committee leadership (51%), National governments (45%), Private citizens (33%), European commission (10%).

These results prove the strong influencing power of organised interest groups in decision-making. The frequency of access of these groups to decision makers is indeed impressive: the great majority of MEPs appear to receive information mainly from interests groups and lobbyists. Moreover, the influence of interests groups in the policy outcome is evident by their direct involvement in the voting process: they are the ones providing more frequently voting instructions to MEPs than either the party leadership or the national governments. Adding to that the frequency of voting instructions by private citizens (33%), which could also be considered lobbyists, one could be allowed to conclude that what comes under the term "lobbyism" is the single most decisive influence in politics.

It is nevertheless comforting to note that there appears to be a balanced representation between opposing "interests" forces. Environmental and consumer groups seem to have equal, if not more, access to policy makers in relation to trade and industry organisations. This does not though denote the influencing power of these lobby groups; which group influences more the eventual vote is difficult to tell. In any case, the overall voting behaviour in the European Parliament since 1979 normally follows well-established political ideologies and party lines (Hix et al. 2002).

2.1
National differences in lobbying activity

The survey data allow for interesting national comparisons and the three big EU Member States, Germany, UK and France provide a good case study: They are the three biggest Member States, have the most voting power in the Parliament and they also have the highest expenditure on S&T in the EU. Nevertheless, their approaches to decision-making are different and this accounts for many contradictions and even frictions in efforts to solve common problems. The cultural differences permeating many social and political issues, should also be evident in the issue of organised-interests influence in decision-making. The survey country breakdown highlights this aspect:

– UK and Germany are the two EU countries with the most frequent contacts to organised interest groups and lobbyists: 73% of German MEPs and 75% of UK ones are having contacts with organised interest groups at least once a week, compared to only 30% of French MEPs.
– UK MEPs are also by far the most open to contacts with lobbyists, with 84% having weekly contacts. Germans are again second from the big countries (48%), while only 5% of French MEPs appear to have such regular contacts with lobbyists.

The opposite seems to hold true for contacts with consumer and environmental groups. In this case, it is French MEPs that have more regular contacts than Germans or British: 24% of French have weekly meetings with environmental organisations compared to 7% of Germans and 15% of British.

In terms of contacts with industry and trade associations, UK MEPs are the ones that have by far the closest contacts; 33% of UK MEPs have weekly contacts compared to 5% of French and 7% of Germans.

In the area of susceptibility to voting instructions, we also see some significant differences. German MEPs receive voting recommendations from interest groups more frequently than other nations. 58% of German MEPs receive voting instructions on almost every vote, compared to 39% of French and 40% of UK MEPs.

It is evident that there are indeed great differences in the way interest groups might influence the decision-making of these countries' politicians. British and German MEPs appear far more susceptible to external influence than French MEPs. There is also an apparent great imbalance on the type of interest group influence, with industry and trade interest groups finding easier access to UK than to German or French policy-makers. Such national differences are very important in the work of lobbying groups and show the complexity of lobbying activity in the new supranational stage deriving from the creation of the European Union (Jauss 2002; Eising 2002).

3
Lobbying in action: case studies from the European Parliament

Based on the Matrix of "TA roles" developed by the TAMI group (see chapter "Towards a Framework for Assessing the Impact of Technology Assessment") one can attempt to identify a set of roles that organised interests fulfil in their ordinary functions within the decision-making procedures. Not all roles of TA map exactly onto those of organised interests and each issue at stake provides naturally unique parameters that delineate the exact overlap between official TA activities and those of organised interests. The most descriptive and direct way to look into this issue is by analysing specific case studies that offer good examples of lobbying activities. Following are four such cases taken from a variety of S&T policy-making activities at the European Parliament.

3.1
Case I: patentability of computer-implemented inventions (European Parliament 2003)

The first case refers to the issue of the patentability of computer-implemented inventions and highlights the "TA roles" of the impact Matrix that are related to "scientific assessment" and "mediation" aspects of TA:

- Patent protection in the European Union is currently provided by national patent systems and the European patent system under the terms of the Munich Convention of 1973. European patents are granted by the European Patent Office (EPO), but are subject to national rules thereafter. In March 2003 the Council of Ministers reached agreement on a common political approach with respect to the

Community patent, covering the principles of the jurisdictional system (based on a unitary court attached to the European Court of Justice), the language regime, cost and fee issues and the role of national patent offices.

- 15% of all applications currently received by EPO aim at the protection of "computer-implemented inventions"; these are inventions which involve the use of software (albeit not computer programmes as such) and include such devices with built-in control software as mobile phones, intelligent household appliances and engine control devices. The number of applications for such patents has risen more three-fold since 1995. Computer-implemented inventions are the subject of a Commission proposal for a Directive on the patentability of such inventions, currently (July 2003) debated by the European Parliament.

A consultation of Member States and the public at large launched by the Commission in 2000, in the context of the ongoing broader consultation on the patent system in Europe, produced a clear demand for harmonisation action at Community level. The same consultation revealed, however, a clear division between the proponents of unlimited patentability and the advocates of open source software, along the lines of *Linux*, the free Unix-type operating system originally created by Linus Torvalds and other software developers. The former were led by companies, such as Microsoft, and sectoral industry associations, such as UNICE (Union of Industrial and Employers' Confederations of Europe), EICTA (European ICT Industry Association), EUSIDIC (European Association of Information Services) and AFEP-AGREF (Association Française des Entreprises Privées – Association des Grandes Entreprises Françaises). The latter camp comprised a range of views, from no patents for software at all to no patents for software running on general-purpose computers. Concerns were also expressed about the capacity of small and medium-sized enterprises to cope with the complex, long and costly patent application procedure. Although the advocates of open standards were by far more numerous in responding to the Commission's invitation, large software companies were favoured by their superior economic weight in the form of employment and investment capacity.

With the proposed Directive the Commission tried to strike a balance between the two camps. The European Parliament's rapporteur, Arlene McCarthy, argued in favour of the Commission's proposal – up to some possible amendments – by stressing that, if it were rejected, "the European Patent Office and its Boards of Appeal would remain the principal arbitrators of the law and there would be nothing to prevent a gradual drift towards the patentability of business methods and the like, as has been witnessed in the US". The Parliament's Committee on Legal Affairs and the Internal Market voted her report in June 2003. While approving the idea of patentability for computer-implemented inventions in general, committee members held widely divergent views. A majority of them agreed on amendments seeking to improve the text of the proposal, especially as regards the definition of "patentable inventions", by clarifying for example when algorithms can be incorporated in patentable inventions.

The first reading of the European Parliament on 24 September 2003 confirmed most of the committee's amendments. However, the plenary voted also in favour of a series of additional or modified amendments, which sought to further tighten the scope of patent protection. Thus it postulated that "a computer-implemented invention may be claimed only as a product, that is as a programmed device, or as a technical production process", where "only" constitutes a non-trivial addition to the initial text. The plenary also adopted a modified version of a committee amendment, which significantly widens the potential for using a "patented technique" wherever this use "is needed for a significant purpose". Whereas the EuroLinux Alliance spoke of "amendments that clearly restate the non-patentability of programming and business logic, and uphold freedom of publication and interoperation", the chief executive officers of Alcatel, Ericsson, Nokia, Philips and Siemens, in a letter to the Council Presidency and the Commission, expressed their concern about what they describe as a sudden, dramatic and unexpected change in the legal climate in Europe if the Parliament's amendments were to be heeded.

The matter is certainly not closed yet, in view of the further procedural steps (Council Common position, EP second reading etc.) that are still pending, but this case illustrates well the activities of organised interests relating to functions of TA. On one hand, direct consultation of the affected parties helped map out the conflict structure that has been described in the Matrix as "Social Mapping"; the process initiated by the Commission gave the possibility to private citizens, companies and lobbyists of all kinds to express their views and the two main camps became soon evident. On the other hand, basic scientific information and analysis of consequences of actions was also provided directly by lobbyists in their effort to influence the outcome of the debate; this is the realm of TA's functions listed in the Matrix under "Scientific Assessment" and considered among the basic functions of TA. Finally, as the outcome is not known at this stage, there is a clear need for mediation and bridge-building that fall in the realm of Participatory TA functions described in the Matrix as "Mediation".

3.2
Case II: restrictions in the cross-border transfer of personal data (European Parliament 2003)

The next case highlights the function of organised interests as "agenda setters" in a process that leads to direct legislation:

- In October 1995 the European Parliament and the Council adopted Directive 95/46/EC "on the protection of individuals with regard to the processing of personal data". In the sense of this Directive, "processing" meant such operations "as collection, recording, organisation, storage, adaptation or alteration, retrieval, consultation, use, disclosure by transmission, dissemination or otherwise making available, alignment or combination, blocking, erasure or destruction" of personal data. The Directive was to be implemented within three years. Chapter IV (Articles 25 and 26) was dedicated to "Transfer of personal data to third countries". Article 25 required that "the third country in question ensures an adequate level of protection". The interpretation of the word "adequate" was at the origin of an intense Trans-Atlantic debate that was to last five years.
- European-based businesses expressed soon their concern about the possible differences that would arise between Member States as to the interpretation of the Directive, as well as about the potential cost of compliance, and argued in favour

of a self-regulatory system. By the end of 1997 the American government became aware of the concerns of its own companies and expressed its opposition to a "comprehensive system of regulation", preferring instead "a system of specific laws targeted to prevent specific abuses", in the words of then US Under-Secretary for International Trade, David Aaron. The issue was soon on the agenda of the Trans-Atlantic Business Dialogue and the semi-annual EU-US summits.

Then, a few days after the implementation deadline and while only four Member States had implemented it, the US Department of Commerce introduced the concept of "a *safe harbour* for US companies that choose voluntarily to adhere to certain privacy principles". American industry was generally supportive of the concept, but consumer organisations were concerned about inadequate enforcement and redress provisions. In March 2000 the European Commission and the US government reached agreement on the outstanding issues of enforcement within the US and integration of US legislation into the system. Under the agreement, companies responsible for serious non-compliance would no longer enjoy the benefits of Safe Harbour.

While the Council endorsed the agreement, the European Parliament did not, despite intense lobbying from the Commission, Member States, the US government and US Congress and international business – European industry and NGOs were less active at this stage. The majority of MEPs was not convinced the Safe Harbour provided adequate data protection. In a resolution adopted on 5 July 2000, the Parliament:

- pointed to the loopholes contained in the "safe harbour privacy principles" proposed by the US Department of Commerce (as regards, for example, the kinds of firms covered, exceptions for publicly available data, ambiguous terminology, the right of personal appeal and the possibility to obtain compensation for individual damage suffered);
- took the view that "the free movement of data cannot be authorised until all the components of the safe harbour system are operational and the United States authorities have informed the Commission that these conditions have been fulfilled";
- regretted that "in the course of the last two years, there [had] been no consultation of European undertakings with regard to the risk of discrimination in relation to US undertakings" and, "in contrast with the US authorities' consultation of NGOs active in the field of consumer protection, the Commission [had] not embarked on any such consultation of European NGOs".

Notwithstanding the Parliament's scepticism, the Commission adopted its Decision 520/2000/EC of 26 July 2000, which permitted the Safe Harbour to be implemented from 1 November 2000. According to a Commission Staff Working Paper published in February 2002 (SEC(2002) 196), 129 US organisations were involved in the Safe Harbour arrangement as of 1 December 2001, but "a substantial number of [these] organisations do not seem to be observing the expected degree of transparency as regards their overall commitment or as regards the contents of their privacy policies". The Commission would therefore "continue to co-operate with the Department of Commerce in encouraging US organisations to join and to insist on a rigorous respect for the transparency requirements of the Safe Harbour".

This is an example of an issue being brought to the attention of authorities (in this case, the European institutions) by lobbying (European and American businesses, in the beginning, the US government and Congress later on). The effect of the lobbying activity was to set the agenda of the policy debate, a similar function to that described in the TA roles Matrix under "Agenda Setting". The debate phase lasted long, due to different cultural and legal traditions (comprehensive regulation vs. self-regulation), which in the case of many participants in the debate were not overcome until the end. Nevertheless, the process bore fruit, with the European side accepting a voluntary instrument and the American side ensuring its rigorous enforcement. The result was Commission Decision 520/2000/EC, which made the application of the Safe Harbour possible. In TA terminology, the "policy aspects" in the "issue" dimension of the Matrix describe the TA functions that are related to such a process, and in particular the roles "Policy Analysis" and "Decision Taken".

3.3
Case III: global warming – The point of view of industry and environmental NGOs (Pedler 2002)

Another case of lobbying activities in S&T related policies, derives from the ongoing debate on "global warming". The debate is rather complex with conflicting views not only from the traditionally opposing sides but also within each side. In this case study we shall briefly describe two different points of view, that of Lafarge, the world's largest producer of construction materials (cement, aggregates and concrete, roofing and gypsum), and that of WWF (the acronym used by the network since 2001 to avoid the confusion between "World Wide Fund For Nature" and the name used until 1986 "World Wildlife Fund").

Lafarge is, like the rest of the cement industry, a major contributor to industrial greenhouse gas emissions, 40% of which come from burning fossil fuels and 60% from the decarbonisation process transforming limestone ($CaCO_3$) to CaO. The company is keen to point out that it is "firmly committed to being proactive on climate change", recognising that "global warming is a phenomenon for which the application of the precautionary principle is justified". Among other things, the company:

- opposes *energy taxes* on energy-intensive (e.g. cement) industry as inefficient and unfair;
- regards *negotiated commitments* as the most efficient means of reducing industrial emissions;
- regards *emissions trading* as a way of reducing the cost of compliance;
- insists on a fair distribution of the burden of compliance among different sectors.

This position contrasts sharply with that of the leading US oil companies, which are adamantly opposed to change. Lafarge, as well as European-based oil companies, take the environmental arguments more seriously. Lafarge, in particular, believes that it has "to prepare itself to be operating in a carbon-constrained world in the medium term". As far as its lobbying strategy is concerned, Lafarge believes, while concentrating on debates and decisions taken in Brussels, public affairs workers should not lose sight of national governments, nor neglect relations with NGOs and academics. It further believes that industry:

– can put forward its point of view more efficiently if all/many sectors are able to agree on common positions;
– should enter the debate early and try to influence it by being "perceived to be part of the solution" (hence the choice in favour of a proactive over a reactive approach);
– pursue dialogue with and be receptive to the ideas of other stakeholders, with all cultural adjustments that this may imply (hence Lafarge's partnership with WWF).

Climate change is one of the priority areas (along with forest protection, fresh-water conservation, conservation of oceans and coasts, species conservation and reduction/elimination of toxic chemicals) upon which WWF is currently concen-trating its efforts. WWF has been active in this field, especially from the time of the first Conference of the Parties (COP1) to the Framework Convention on Climate Change (UNFCCC) agreed in Rio in 1992. It was COP1 (Berlin, March 1995) that set the ground for the Kyoto Protocol to be concluded 2? years later.

WWF has built up its lobbying presence in Brussels following the opening of a European Policy Office in 1989. In relation to climate change WWF's lobbying efforts concentrated on:

– keeping "EU policy-makers focused on domestic policies and measures";
– ensuring that the EU "would take an environmentally progressive position con-cerning the rules for ... the Clean Development Mechanism" planned under the UNFCCC.

In the years preceding COP3 in Kyoto the WWF position was that greenhouse gas reduction targets to be agreed upon "should be based on sound policies and measures", which "should clearly demonstrate what can be done economically and technically". WWF backed their position with a study commissioned from the University of Utrecht, which concluded that a 14% reduction in CO_2 by 2005 was fea-sible for the EU – in Kyoto the EU proposed a 15% reduction in CO_2, CH_4 and N_2O emissions by 2010. WWF believes that the study was influential in setting the scene for a series of energy policy measures recently decided or under discussion at EU level (a 12% target for the share of renewable energy sources in electricity genera-tion by 2010, increasing cogeneration, efficient heating of buildings and progres-sive standards for appliances).

In WWF's assessment of the outcome of COP3, where industrialised countries agreed to cut greenhouse gas emissions, averaged over the period 2008 – 2012, by 5% compared to 1990 levels, "efforts aimed at getting industrialised nations to agree on absolute emission-reduction targets in Kyoto were only successful because the EU had backed their proposal with policies and measures which were reason-able and were widely supported by WWF and other NGOs".

We have illustrated in this section the considerable efforts made by two different and often (but not always) opposed interest groups to set the agenda of the world-wide climate debate. Their efforts were proportionate to the stakes in this debate, seen either from the economic point of view or from the point of view of ensuring a sustainable world for future generations. Despite the scientific uncertainties and disagreements that have characterised the climate debate, the basis of the ultimate decisions has to rely on scientific facts, coupled with the political will to apply the precautionary principle – according to Article 3(3) of the UNFCCC, the Parties "should take precautionary measures to anticipate, prevent or minimize the causes of climate change and mitigate its adverse effects. Where there are threats of serious or irreversible damage, lack of full scientific certainty should not be used as a reason for postponing such measures ...".

TA can in such cases work closely with interest groups in a mutual albeit critical exchange of information, both about the scientific facts and the political dynamics of the situation (described in the Matrix as "Scientific Assessment" and "Agenda setting"). Given the number and complexity of the different actors involved in such a global debate, understanding the structure of conflicts (Matrix: "Social Mapping") is a valuable ingredient in the analysis of interest groups and TA certainly has valuable expertise to apply in this area. The results of such an analysis can then be implemented for the purposes of mediation (Matrix: "Mediation"), although the role of TA in this case is *de facto* limited in scale. It can become indispensable, however, when it comes to overcoming conflicts within individual countries concerning the implementation of internationally agreed measures. In Europe TA may also have a limited supranational impact given the particular role played by the EU in the negotiations and the increasing EU competence in the energy sector.

3.4
Case IV: air pollution; car emissions and motor fuels (Pedler 2002)

The final case study deriving from the debate on car emissions and motor fuels in connection with air pollution illustrates the activities of interest groups that permeate all major functions of TA as described in the Matrix. The long legislative process (codecision) that led to the adoption by the European Parliament and the Council of Directives 98/69/EC on measures against "air pollution by emissions from motor vehicles" and 98/70/EC on "the quality of petrol and diesel fuels" provides a good example of the different conflicting interests (sometimes among players in the same camp) and varying alliances on an issue that relies heavily on scientific support.

The first limit values for CO and unburned hydrocarbon emissions at European level were introduced by Directive 70/220/EEC. Subsequent directives successively reduced the limit values, added limit values for nitrogen oxides (NO_x) and particulate pollutants from diesel engines, extended the standards to all types of passenger cars and introduced requirements for checking the conformity of production. Article 4 of Directive 94/12/EC required from the Commission to propose standards to be enforced after the year 2000; besides the tightening of car emission standards, the Commission proposal was expected to include complementary measures on improving fuel quality. The Commission proposals mentioned in the beginning were submitted to the Council and the Parliament in 1996 and the final acts (following the two readings of the codecision procedure and conciliation) were adopted in October 1998.

Following the 1994 Directive, which introduced more stringent values, the automobile industry lobbied the institutions requiring full participation in the development of further measures. Then the Commission launched the so-called Auto-Oil research programme in collaboration with the car manufacturing industry, represented by the European Automobile Manufacturers Association (ACEA), and the oil industry, represented by EUROPIA. The Commission co-funded the programme, whereas industry participated in the research, making use of its own resources and technical means. The programme resulted in the Commission proposals of 1996.

Interestingly the interests of the car industry were to a large extent opposed to those of the oil industry insofar as they were supposed to share, each from his side, the burden of the emission limitation process (car exhausts vs. cleaner fuel). They found however common ground in demanding and achieving that the Commission proposals rely on "cost-effectiveness", and not on Best Available Technology (BAT) as demanded by NGOs. The latter, although not involved in the Auto-Oil programme, lobbied actively the European institutions during the process that led to the adoption of the two directives. It is indicative of a developing trend that NGOs were invited to participate as "stakeholders", alongside the major industries, in the launch of a new programme of consultation and development ("Auto-Oil II") that should result in proposals for new, more stringent air quality targets for 2010.

If one looks at the process leading up to the directives of 1998, it appears that the automobile industry was successful in introducing, through the Auto-Oil programme limits acceptable to its members. In the conciliation process, the European Parliament accepted the limits proposed by the Commission and the Council in its common position, obtaining in return the 'mandatory' character of the medium-term limit values (initially intended as "targets") for 2005. The Parliament also obtained the provision that "on-board diagnostics (OBD) should be introduced with a view to permitting an immediate detection of failure of anti-pollution vehicle equipment".

By contrast, in the case of the directive on the quality of fuels, the limits adopted were distinctly lower than those proposed by the Commission, although they largely remained at the levels of the Council's common position and did not go as far as the Parliament would have wanted. In exchange, the limits were mandatory for 2000, but not for 2005.

Finally, as far as the NGOs are concerned, they did not succeed introducing BAT as the basis of the legislation, nor did they manage to reduce the car exhaust limits, but obtained tighter limits for fuels and, more importantly, established themselves as full participants in subsequent steps.

This case illustrates well the current trends in European policy-making, which largely reflect the increasing role played by interest groups in this process. Lobbying activities succeeded in providing a comprehensive "Scientific Assessment", then promoted "Agenda Setting" and were eventually reflected in the "Decision Taken": all representing main functions of TA as described in the Matrix. It also illustrates the widening area of EU competence which, coupled (especially after the entry into force of the Amsterdam Treaty) with the increased application of the codecision procedure, raises the European Parliament to the level of co-legislator on the same footing as the Council. This implies an ever-widening area of activity for lobbying, but also an ever widening number and variety of actors to be lobbied (Van Schedelen 2002; Lahusen 2002).

4
Conclusions: lobbying and TA – synergies and conflicts

In the description of the case studies and the analysis of the MEP survey above we have seen again that it is a fact of contemporary life that lobbyists play an important role as information providers and influencing factors in political decision-making. When their activity focuses on technical issues, then their ways inevitably cross with those of TA professionals and all sorts of technical experts advising politicians. In certain circumstances these encounters may be compatible and mutually reinforcing, in others they may be conflicting and mutually destructive. This is a very dynamic relation and may change from one type to the other between the same people and institutions and for similar issues or within the same debate, depending on a host of very complex and not necessarily mutually independent parameters. This calls in turn for continuous vigilance and preparedness to face unexpected developments.

The main question that lobbyists and TA professionals face alike is how they can best *manage* their complex and variable relation. With appropriate management the unquestionable expertise of many lobbyists complements or can complement one or more of the TA roles that we identified in the impact Matrix. TA professionals and institutes are therefore well-advised to include a good sample of individual lobbyists and lobbying organisations in the list of people they keep an eye on and occasionally converse with. Hopefully this attitude will be reciprocal, as lobbyists have as much to gain from good TA experts, as the latter have to gain from good and knowledgeable lobbyists.

From the point of view of a TA project, lobbying should be seen as a multi-faceted activity: It is often an inevitable *external parameter* to a particular situation, with appropriate management, however, can be incorporated in the project as an *internal parameter*. Treating lobbying as a parameter is helpful when no big variations in its practices are expected over the time span of interest, as is the case for mature fields, not currently experiencing revolutionary change. Then it may be sufficient to follow lobbying activities via news releases, position papers, the press and mass media, and the Internet. As an external parameter it may have a constructive/amplifying effect or a destructive effect from the point of view of the balanced description of the situation pursued by the TA project. Incorporated as an internal parameter in the TA project, lobbying can be critically tapped for its useful elements, while discarding those of its aspects that would be detrimental to the integrity of the TA project (see also the concept of "situational appreciation" developed in the chapter "The Practice of TA – Science, Interaction, and Communication").

As lobbyists are generally active or ready to jump into action on short notice, and their activity usually has an unquestionable influence on the actual situation, a more dynamic approach, treating them as actors in any confrontation of (actually or seemingly) opposing interests is usually more appropriate. To start with, they may be regarded as *external actors* who try to influence the situation with a very specific bias and pursue a desired outcome at any cost. As such, they may well be perceived as counter-productive from the point of view of TA's efforts to establish a factual and impartial picture of the particular confrontation. This may be occasionally inevitable, depending on the character of the confrontation and the attitudes of those involved in it. In general, however, TA can turn this around and look at lobbyists as a resource to be tapped. For rapidly evolving and complex technical issues, where the available expertise can never be sufficient for a thorough appreciation of the situation, technically literate lobbyists can be part of the limited pool of scarce knowledge that should by no means be wasted.

It is again a question of appropriate management of the dynamics of the situation by TA experts, so that lobbyists can be co-opted as *internal actors*, who contribute to the acquisition of the necessary knowledge and, depending on the particular configuration of positions, can even act as amplifying factors in the task of TA. This does not mean that lobbyists will be asked to denounce their allegiance to those they lobby for and act against their natural interests. What they offer will have to be made use of with a critical mind, as it is likely to contain lots of useful facts and insights. Neither does the benefit need to be one-way. The lobbyists themselves have a lot to gain from the expertise of TA practitioners and the opportunities to meet and possibly influence other actors in the particular situation during the activities organised by the latter. Knowing that TA is by its very nature non-confrontational, lobbyists can look at TA products and activities as a valuable resource which they are free to and, in fact, should constructively criticise when it does not live up to its purported standards.

In conclusion, given their inevitable parallel existence, TA should orient itself towards the appropriate management strategy that will allow it to utilise the useful elements that lobbying has to offer. Whether the issue is keeping abreast of complex technical information, mapping social conflicts, or reviewing relevant policies, there are surely opportunities for both lobbyists and TA experts of constructive exchanges to mutual benefit.

References

Bouwen P (2002) Corporate Lobbying in the European Union: the Logic of Access. Working paper for "Interessendurchsetzung im Mehrebenensystem", DFG, Mannheim 4–5 July

Eising R (2002) Die Europäisierung der deutschen Interessenvermittlung und Interessengruppen. Working paper for "Interessendurchsetzung im Mehrebenensystem", DFG, Mannheim 4–5 July

European Parliament (2003) Lobbying in the European Union: Current rules and practices, Working document, Directorate General for Research, January 2003

Grande E (2001) Institutions and Interests: Interest Groups in the European System of Multi-Level Governance. Working paper for the ECSA Seventh Biennial International Conference, Mai 31, 2000, Madison, Wisc.

Hix S, Noury A, Roland G (2002) How MEPs vote. Press release, London School of Economics and Political Science, 21 September 2002

Hix S (1999) The Political System of the European Union. St. Martins Press, USA

Jauss C (2002) Aktuelle Entwicklungen in der Struktur der europäischen Interessenvermittlungslandschaft. Working paper for "Interessendurchsetzung im Mehrebenensystem", DFG, Mannheim 4–5

Kohler-Koch B (1997) Organized Interests in the EC and the European Parliament. European Integration online Papers, 1, 9, 1–22

Lahusen C (2002) Professional Consultancies in the European Union: Findings of a survey on commercial interest intermediation. Working paper for "Interessendurchsetzung im Mehrebenensystem", DFG, Mannheim 4–5

Pedler R (2002) European Union Lobbying – Changes in the arena, edited by Robin Pedler. Palgrave, UK

Van Schendelen R (2002) Machiavelli in Brussels: The Art of Lobbying the EU. Amsterdam University Press, The Netherlands

Industry Technology Assessment: Opportunities and Challenges for Partnership

Robin Fears and Susanne Stephan

This paper draws on experience in the UK pharmaceutical sector to describe a case study on industry approaches to Technology Assessment (TA). Before providing this detail, a general introductory section reviews the European background to TA, identifies some industry trends (with particular reference to experience in Germany) and provides a taxonomy to compare and contrast the public and private sector approaches to TA.

1
Introduction

1.1
History and classification of TA

The current era of TA first emerged in the USA in the 1960s, prompted by perceptions of deficiencies in communication between the executive and legislative branches of government. The establishment of the first US TA institute (OTA) in 1973, with the objective to support and consult with Congress on technical issues was followed, in the 1980s, by TA offices in the EU Member States – France, the Netherlands, Denmark, UK and Germany – and the European Parliament (Salo & Kuusi, 2001). These European developments were responding, in part, to concerns that advances in science and technology (S&T) were insufficiently controlled, but the institutional process to conduct TA varied greatly across the Member States, differing according to the elements that each country considered central to S&T advice. Grunwald (2002) categorised TA according to the following parameters:

- *Tasks and functions* – the most important of these relate to policy support and technology-forecasting (early warning system) as well as covering problem analysis and problem solution in the societal context;

- *Methods* – these differ between TA institutions and countries, the most well known being participatory TA, constructive TA, rational TA, Innovation and Technology Analysis (ITA) and product TA. The latter will be exemplified later in this paper;

- *Fields of action* – TA has been used in many different research areas, it is clearly important to identify the concrete research activities that are to be assessed;

- *Target groups* – this issue has been contentious. Should TA address all sections of society or focus, in particular, on the policy-making community? Is industry a target of TA or a partner?

There is a paradox (Salo & Kuusi 2001), exemplified by the US Office of Technology Assessment, in that parliamentary institutions may fall out of favour with legislators, even if their work is appreciated elsewhere. In order to resonate with parliamentary needs, TA outputs must be timely and credible. Other variants of TA – such as integrative, interactive – have been formulated (Bröchler et al. 1999) as part of the postulated relocation of TA from the domain of policy advice to that of research, innovation and enterprise. Hennen (1999) has critically reviewed the claims of those calling for a fundamental re-orientation of TA to address perceived deficits in its policy advisory role.

Given the variation in the principles and practice of TA, the EU TAMI project has formulated a common definition that would characterise most European TA approaches: "…a scientific and communication process with the aim of contributing to rational public and political opinion formation on societal aspects of science and technology". Central to this paper is the debate on whether industry can be a collaborator in TA (further discussion by the German Institute for Technology Assessment and System Analysis is on www.itas.fzk.de).

1.2
Considerations in favour of industry as TA collaborator

In the EU, approximately three-quarters of all research is funded and conducted by the private sector – most major innovations are made by industry. Furthermore, an increasing pace of technological innovation has led to a greater dependence on the contribution made by leading industry sectors. Thus, national TA activity cannot afford to exclude the most important source of innovation, it has to develop procedures to interact with industry and include industry perspectives. It has often been suggested that classical TA is inevitably too late with its efforts to analyse the (unintended) impacts of technological developments; while Hennen (1999) constructs a robust defence of the timeliness of public TA, he also recognises the importance of embedding TA as a concept more deeply within the innovation processes in firms and public research institutions.

The growing importance of the debate on TA-industry collaboration is illustrated by several German initiatives:

– The BMBF (German Ministry for Education and Research) has introduced Innovation and Technology Analysis, focusing on the relationship between industry and TA (www.idta.de);
– The university "WHU" (Otto Beisham Graduate School of Management) published "Technology Assessment – A management perspective" (Weber et al. 1999) in which TA was evaluated from the perspective of industry consultancy and calling for more competition in the TA practitioner community;
– The TA working group, Northrhine-Westphalia (AKTAB) addresses TA in and for industry in the concept "innovation-oriented TA" (Steinmuller et al 1999).

The counter-argument to productive and balanced TA-industry collaboration asserts that industry interests reside only in economic factors and shareholder value. However, industry cannot afford to ignore the side-effects/unintended impact of its products (particularly in times of increasing competition) and industry is expected to ensure and maintain trust among consumers. Arguably, innovation-oriented TA has the goal of establishing processes of societal self-control through the consideration of social demands in industry as an alternative to state intervention in the field of technology policy (Ropohl 1999; Hennen 1999). Again using the example of Germany, socially-responsible industrial innovation is to be found in major companies such as BASF and Daimler-Chrysler (Becks and Gelbke 2001; Minx 2001).

The issue as to how (and why) public sector TA and industry form collaborations requires further work (and the example of the pharmaceutical industry is considered subsequently). In some cases it may be a short-term strategy for impact during the legislative process, in other cases it may become a permanent partnership for lobbying purposes. Whatever the provenance, TA-industry partnership might be one instrument to create awareness in public and policy-making areas, and thus to increase the power of TA.

1.3
Considerations against industry as collaborator

TA credentials of independence and neutrality are two of its most important aspects. Most TA institutions are funded by public money and must avoid commercial conflict of interests. TA is also expected to deliver within the overall framework of common welfare where social and environmental questions are important, but there is usually less commitment by private companies to examine the consequences of their products in a more general context. Furthermore, national level TA is assumed, democratically, to represent a plurality of stakeholder opinions – industry can only be one of these stakeholders.

Space does not now permit a full discussion of the different opinions about the effectiveness and merits of collaboration between public TA bodies and industry. However, in summary, and as the final step in the introduction to the specific case study that follows, a general taxonomy of differentiating features can be attempted.

The following case history develops the theme of complementary activities and partnership – each constituency needs to be aware of the other's interests and goals and awareness is a major step towards successful collaboration. Hennen (1999) has concluded that the opportunities for innovation-oriented TA are, essentially, confined to pre-competitive areas (at the interface between basic and application oriented research) and are linked with exploration of market opportunities rather than a more substantial integration of non-technological interests of society. Is this view now sustainable? By reviewing examples drawn from both emerging technology assessment and product-associated technology assessment, the rest of this paper identifies an evidence base on which further judgement about TA can be attempted.

2
Impact of new technology in health care

The increase in total spending on health care across Europe suggests that health care services are becoming more expensive. Is this spending productive – are services delivering increased health gain? Technology is one of the most important driv-

Figure 1: Industry versus Public TA

Criteria	Industry TA	Public TA
Common welfare	Research on specific products; "enlightened exploration" of market opportunities but less commitment to more general issues	General assessment of consequences of technological process, including unintended impacts
Constitution	Integrated in internal innovation processes and enterprise interests/industrial practice	Usually public financed or governmental institutions
Independence	Subject to company obligations on shareholder value and economic aspects	Attempts to be independent and neutral
Operability	Key criterion – need for practical solutions	Results are mostly related to political solutions
Customer orientation	High (and variable customer involvement)	Different customer orientation as for Industry TA (Policy maker, authorities, R&D, NGO's, Media, Society in general)
Market orientation	High	None
Timeframe	Often proactive; can be long-term or short-term (as appropriate)	Proactive; mostly long-term
Time pressure	High	Lower than by the Industry
Project management	High	Low
Costs	Influence research volume of the project	Less significant issue

ers of health spending and health economists identify technological change as a primary reason for the increase in the health sector share of GDP (Wanless 2002). Solving the problems of rising health care costs and increasing demands (made, for example, by an ageing population) requires better understanding of the contribution made by new health care technologies in reducing disability and premature mortality, lowering care costs and transforming pharmaceutical R&D processes (Pardes et al. 1999).

The introduction of new health technologies can be dramatically cost-effective. Studies on a range of therapies and procedures – treatment of cataracts, coronary heart disease, depression, low infant birth weight – show that benefits can greatly outweigh costs (Canning 2003).

In a landmark US study, the Lasker Charitable Trust (2000) examined the macroeconomic return for medical research by comparing the total economic value of the reduction in cardiovascular mortality, arising from better survival rates in the aftermath of heart attack and stroke and the reduction in risk factors, with the investment in medical research. The total economic value to the US of the reduction in mortality from cardiovascular disease over the period 1970–1990 averaged $1.5 trillion annually; assuming one-third of the gain could be attributed to medical research, the return was approximately 20 times larger than the spend on research.

Research on cardiovascular disease also exemplifies another key issue for TA impact discussion – the variation in practice among health service providers even when there is a convincing evidence base for benefit. For example, the use of angioplasty with stents to unblock coronary arteries is cost-effective and is recommended by UK advisory bodies. However, in the UK the rates of angioplasty are often lowest in those regions with the highest burden of coronary heart disease and, overall, the UK rate is about one-third of that in Germany (Canning 2003).

In general, the UK is relatively slow to adopt new medical technologies (Wanless 2002). UK spending compares poorly with other industrialised countries: 4.2% of total health care funding is spent on medical technologies compared with the Northern European average of 6.4% (Medical Technology Group 2003). A study of technological change in heart attack care characterised the UK as "late and slow" in adopting new technology, compared with "early and rapid" for the US and "late and rapid" for France (Wanless 2002). The rate of uptake of new drugs in the UK is about half that in Germany and one-third that in France.

Because some new health technologies may be costly and offer few extra benefits, countries must develop processes to identify and select the most cost-effective interventions and then implement them without delay. However, policy-makers may have to take account of criteria other than cost-effectiveness because of health service priorities and societal pressures: health services often prefer to avoid difficult choices even if the result is inefficiency. It is important to avoid considering industry roles and responsibilities in isolation – TA is defined by WHO as encompassing not just equipment, devices and medicines but also clinical procedures and the organisation and support systems within which health care is provided (Wanless 2002).

The extent to which Health Technology Assessment (HTA) has similar characteristics to other TA might be debatable; it is a premise of this paper that HTA has great scope for collaboration, such that private-public sector differences in TA attitudes are reconcilable. The pharmaceutical industry is often viewed as a proving ground for innovation (Martin, 2003) – it attracts attention because of its high policy profile and because its performance appears to be affected directly by public policy. Whatever the generalisability of the conclusions in HTA, there can be little doubt that HTA is, itself, a worthy field of study. In this field, TA has the potential to directly affect both health and wealth creation across the EU, and there is an imperative to provide an evidence base to inform complicated and sensitive policy decisions to be taken about the prioritisation of resources.

3
Health technology assessment

The purpose of HTA, according to OECD criteria (2003), is to facilitate the allocation of resources in relation to the goals of the health care policy makers – by generating/evaluating evidence of the relative effectiveness and costs of the technology and by ensuring that this information is used appropriately. The knowledge generation function includes: identifying the evidence; synthesising health research findings about the effectiveness of different interventions; evaluating the economic implications; and appraising the social, ethical and organisational implications.

As discussed in general terms in the Introduction, the increasing pace of technological innovation in health care inevitably leads to higher dependency on TA performed by industry; company TA may be performed solely "in-house" or in formal partnership with other stakeholders (particularly, regulators, customers, academic groups). While there may be significant differences in form and process between TA conducted in the public and private sectors, health care companies share the general objectives in developing effective and relevant methods in TA and ensuring that the TA has impact on the other stakeholders – patients, health services, regulators and policy makers.

4 Research outputs

The challenge facing health care companies in assessing their R&D/technology/ performance over a shorter timeframe than the product lifecycle) is similar, in some respects, to the issues facing all researchers in identifying the value of their outputs. Metrics for measuring health care research outputs, outcomes and impact cover a wide range of payback indicators (Buxton and Hanney 1998) of relevance both to companies and the public sector:

- Bibliometric indices to measure knowledge generation, citation metrics to measure impact on other researchers and patent-linkage metrics to identify inputs to innovation.
- Case study analysis to link knowledge generation with advances in medical practice (for example, the development of clinical guidelines and new interventions), an approach pioneered by Comroe and Dripps (1976).
- Cost-effectiveness evaluation of individual products and services.
- Economic impact assessment, whether in terms of the contribution by the research enterprise at the workplace level to the firm or to the local economy or to national competitiveness, or at the macro level in terms of the return on investment (Lasker Charitable Trust 2000).

The methodological challenges faced in research, technology and product assessment are compounded if it is desired to identify attribution; for example to distinguish the relative contributions made by industry and academic researchers in the development of novel health care technologies, or the relative contribution made by multiple funding agencies, or the relative contribution made by medical research and other disciplines (such as engineering) in technology convergence.

In considering best practice in research assessment, the example of the US National Institutes of Health in responding to the 1993 Government Performance and Results Act is of particular interest. This Act was intended to enhance the effectiveness, efficiency and accountability of publicly funded research and stimulated the NIH to address strategic questions relating to the nature of its mission, plans to address its goals, measurement of productivity and use of the indicators to make qualitative improvements in performance (NIH, 1999). The results – a set of narratives that are relevant in many aspects to the domain of TA – published on the NIH website essentially provide a history of the major advances in biomedical research innovation. The NIH activity provides a benchmark for how research funders (public or private) can act strategically to demonstrate accountability and good research governance and for how the information gained from TA can be incorporated into the performance standards of the technology provider.

5
Pharmaceutical R&D

The global pharmaceutical industry sector is characterised by high R&D investment. Product development time is lengthy (approximately 10 years from discovery to launch), uncertain and expensive (industry estimates approximately £500 million for a new product, including opportunity costs and failures, although others would suggest a lower net figure). In the UK, pharmaceutical and biotechnology companies dominate R&D activity (DTI 2002 R&D Scoreboard) – contributing 37% of all industry R&D. UK pharmaceutical R&D intensity (investment as a proportion of sales) is 14.6%, compared with 2.2% for all sectors. In the pharmaceutical/biotechnology sector (as for other sectors), those companies committing to above average R&D spend tend to enjoy above average sales growth, productivity and shareholder return. The EU has identified the pharmaceutical/biotechnology sectors as a key area for current and future competitiveness.

For the pharmaceutical sector, and its customers, TA is often interpreted to mean Product Assessment, and competitive advantage to the firm. However, there are also examples of TA relating to those underpinning technologies and shared pharmacology that have contributed to the development of multiple products (for example, the cholesterol-lowering Statins). It is of considerable economic value to a company to be able to predict and then capitalise on enabling technologies that may transform both medical practice and pharmaceutical R&D practice. The example of pharmacogenetics as an emerging technology will be considered later in this paper.

5.1
Product assessment

Because of the nature of the health authorities in most European countries – funded by tax or insurance and with near-monopoly purchasing power – there is considerable interest in controlling the price of medicines and in demonstrating cost-effectiveness. In consequence, there may be tensions between the Member States' and EU policy-maker's desires for supporting industry innovation and competitiveness and for their delivering low cost health services. Historically, this tension has been a cause for concern for companies but there has been recent shared commitment, to finding solutions to support company innovation and health service value, in the collective efforts of Member States, the European Commission, companies, patient groups, health services, insurers (UK Pharmaceutical Industry Competitiveness Task Force 2001; European G10 Medicines Group 2002).

It is not the purpose of this paper to provide a detailed account of pharmaceutical company operations; the complex nature of the processes whereby medicines are evaluated for safety and efficacy, the ethical and legal considerations, the importance of consulting the public and the management of delivery within the health services are comprehensively described elsewhere (for example, Royal College of Physicians Report 2000). In seeking to demonstrate both innovativeness and value-for-money in novel health care products and services, companies will engage in a significant amount of activity to measure cost-effectiveness (as well as safety and efficacy) during their clinical R&D programmes. National regulatory and advisory bodies are also increasingly building cost-effectiveness Product Assessment criteria into their registration, pricing and formulary decision-making. Space does not now permit a detailed description of the typical methodology undertaken by companies – in conjunction with other stakeholders – to measure outcomes (see, for example, Vaccani et al. 1997); the UK Office of Health Economics (www.ohe.org) provides a comprehensive Health Economics Evaluation Database of studies of cost-effectiveness and other forms of economic evaluation of medicines, other treatments and medical interventions.

It requires a large amount of clinical R&D, followed by evaluation of product impact in routine use, to characterise the value of a novel intervention. Thus, it is important for industry and regulators to agree on the processes of Product Assessment – and to establish what data are sufficient to allow a product onto the market, to be followed by assessment in use. There are issues not just for the methodology and timing of the assessment process but also for the nature of the evidence to be assessed, for example the acceptability of surrogate endpoints as an alternative to mortality statistics. There is a need for rigorous and independent assessment and validation of the surrogate and biomarker indices used. There is also growing opportunity for EU health service systems to provide patient informatics databases for shared (public and private sector) use to monitor health outcomes and assess the impact of therapeutic interventions (Fears and Poste 1999).

TA for new biotechnology products in health care can be a fatal flaw in a biotech firm's business model – an over-optimistic revenue prediction by the firm may, for example, under-estimate the importance of receiving reimbursement status for the product. In recent cases (Bouchier 2003), for example, it was observed that companies did not properly assess the cost-effectiveness of their mass-produced engineered skin products. As a result, coding and reimbursement authorities approved reimbursement for these products for a much smaller market than the firms had anticipated, forcing them to file for bankruptcy at the end of 2002. The reality is that authorities approve and reimburse products based on what is most effective and cost-effective for a particular indication – their decisions are not dependent on how scientifically innovative the technology is behind the therapy.

5.2
Merging technologies in product development – need for informed, innovative purchaser

While it is self-evident that products must be assessed according to the value that they deliver, it is also vitally important that health services and their regulators develop a coherent strategy to assess and realise the substantial benefits that new technologies can confer. Much has been written about the challenges and opportunities for improving the quality of health care that are afforded by advances in genomics and molecular medicine (Richards 1999; Fears et al. 2000). There is room for a pluralistic approach with commitment to public-private partnership in strategic thinking on TA, to inform health services priorities and preparedness. Health services will have to exert a market pull by acting as an informed and innovative customer.

The imperative for public policy-makers to support innovation in the general context is reinforced (Salmakaita and Sala 2002) by the need to avoid the "anticipatory myopia" that arises from insufficient activity to evaluate future opportunities, lack of incentives to share information and structural inertia that inhibits adoption. What is necessary, therefore, is not just partnership in TA but also in Technology Policy Assessment – shared commitment to a set of activities that includes developing the evidence base, acting with transparency in using information, generalising and learning from previous experience, piloting new actions and evaluating their impact, identifying performance metrics and analysing net effectiveness (Salmakaita and Sala 2002).

6
UK health technology assessment mechanisms

The body of evidence cited previously (Wanless 2002) showed that the UK lags behind other EU countries in the use of technology in the health care sector. In addressing this under-performance issue, it has been broadly agreed that the optimal timing, speed and extent of technology diffusion should depend on the expected impact on health outcomes; the appropriate response to new technologies is for rapid and consistent diffusion across the health services once robust evidence of cost-effectiveness is available.

The UK has initiated a Health Technology Assessment Programme for the National Health Service (www.ncchta.org), covering screening, diagnosis, treatment and rehabilitation; measuring effectiveness, cost-effectiveness and impact. This appraisal process is based on multidisciplinary teams using randomised clinical trial data, generating evidence for policy and practice. The UK provides an interesting example of how one Government Department (Department of Health) is involved as evaluator, purchaser and user of novel products and technologies and as funder of new technology development in partnership with industry (for example, the Medical Devices programme, www.healthtechnologyportal.org.uk).

Such programmes are strengthening the relationship between research and practice in health care but it is also important to better understand how evidence-based medicine is used within organisations, when subject to individual values in decision-making (Dargie, 2000). Thus, TA cannot be separated from social, environmental and organisational issues, whether in terms of the focus of assessment (for example – individual technology or broader clinical practice) or unit of assessment (for example – patient or population). An illustration of the interest in the UK in understanding the interaction between innovative health technologies and wider changes in society is provided by the Research Councils Programme on Innovative Health Technologies (ESRC 2001). Recent work examined the development and introduction of the artificial hip – in terms of the reception by professionals, patients and policy-makers (and their expectations) and the dynamics of the companies involved in the manufacture and distribution of prostheses – in order to identify models for the study of other innovations and clinical assessments (www2.york.ac.uk/res/iht/projects/1218252045.htm).

6.1
National institute for clinical excellence

The UK National Institute for Clinical Excellence (NICE) now has a central role to play in assessing clinical benefits and costs of health care interventions; appraising drugs and devices as health technologies. NICE was created to advise on best clinical practice to NHS clinicians, to those commissioning NHS services, and the public, on topics chosen to reflect national clinical priorities. The NICE process of TA is described in detail on www.nice.org.uk; NICE provides detailed guidance on the nature of the evidence to be provided by pharmaceutical companies and other manufacturers to TA process by independent experts and a representative appraisal committee. The case history of the evaluation by NICE of one innovative technology – disease modifying drugs (reducing inflammation by unknown mechanisms) to treat multiple sclerosis – is well described by the Parliamentary Office of Science and Technology (Postnote 2002). Industry is a key participant in the TA process, by providing evidence, by nominating clinical and patient experts to provide their perspectives, by commenting on the TA report and the NICE judgement.

Pharmaceutical companies supported the introduction of NICE but have now made recommendations to resolve weaknesses in the process that potentially impede the success of TA in supporting health care advance. Key concerns relate to selection criteria for TA, timing of TA, performance measures for those engaged in TA, policy implications relating to the impact of TA. Resolution of these issues may be generalisable for other TA practices. For example:

- Procedure for selecting products/technologies for review should be open and subject to consultation with all stakeholders.
- Timing of TA should be set so that data on widespread use are available and the criteria for TA need to be developed in discussion with patients. If a new product/technology is assessed prematurely (for example, before market launch) then insufficient data may be available, but also the subsequent market introduction may be delayed and patients denied benefits.
- More emphasis should be placed on cost-effectiveness in terms of outcomes for patients and implications for other Government spending rather than the perceived primary focus on affordability (with implications of rationing).
- Success of NICE depends not just on quality of assessment but also on how the subsequent guidance is implemented – the evidence for impact needs to be monitored, performance indicators agreed and variation in practice reduced. Thus TA must be followed by audit (and sanctions, if appropriate).

For the TA system exemplified by NICE to be effective, and for similar systems to be introduced across the EU, it is essential that industry retains confidence in the process otherwise there is a risk of further migration of R&D investment away from the EU. The performance of NICE as a model for elements of collaboration, reconciling public service and industry interests, merits further study by TA practitioners.

6.2
Risk sharing

The NICE disease-modifying multiple sclerosis drug assessment was innovative, not just because of the nature of the technology reviewed but also because it led to a risk-sharing scheme. This allowed the NHS payment for each drug to be adjusted depending on the drug's performance, so that cost-effectiveness is maintained at an agreed threshold (Miller 2003). Essentially, this is a commitment to continuing TA. It needs substantial investment to generate reliable data but represents a very interesting industry-patient-health services partnership approach to collecting new data while allowing the patient to benefit from new therapies. The risk sharing principle was recently extended to primary care for another major pharmaceutical class- the Statins – whereby the drug manufacturer agreed to refund the health services if failing to meet performance targets when used under appropriate conditions (Chapman et al. 2003). An outcome guarantee has the potential to ensure predictable health gains for a given drug expenditure and this is particularly relevant if concerns about inappropriate use would otherwise result in resistance to uptake. Evaluation of pilot projects will establish the overall acceptability of the partnership model to patients, health services and companies as the stakeholders, and the future applicability in product/technology assessment.

While it is certain that patients should be involved as stakeholders in assessment, it has also been observed that patients may be more willing to take risks than health care professionals. It is critically important, therefore, that TA and product assessment be based on a sufficient evidence base of high quality scientific data that has been peer reviewed and assessed by regulatory authorities and that procedures are in place to allow follow up to measure outcomes. The proposal to use pentosan polysulphate in the treatment of vCJD (www.doh.gov.uk/cjd/pentosan.htm) illustrates the problems precipitated by individual advocacy in the absence of sufficient evidence.

6.3
Capturing unexpected benefits

One other issue for the continuing assessment of new health care products and technologies – and for the role of patients together with company and academic researchers – is the importance of capturing the unexpected benefits of medical research. Among the uncertainties that continue after product launch are those benefits that were unanticipated when the research was performed (Gelijns et al. 1998). Many new indications have been discovered after drugs and devices were introduced into clinical practice and there are several policy implications relating to the need to improve the identification of secondary uses as part of the ongoing TA partnership. For example, it is desirable to build closer communication between the medical disciplines, to create better interdisciplinary communication between academia and industry, to share best practice in devising new initiatives for clinical evaluative research and to identify new incentives for the private sector.

7
Pharmacogenetics

In the OECD work on HTA, previously cited, a prime concern to countries relating to the assessment and diffusion of emerging technologies was the implications arising from the advances in genetics. The impact of genetics on medicine may prove to be greater than that of any other previous scientific advance (Fears et al 2000) and, as in the other HTA areas, there is considerable opportunity to improve the mechanisms for systematically establishing the effectiveness and appropriateness of interventions, and for ensuring that these are adopted consistently (Zimmern and Cook 2000). Human genomics research will have impact on medical practice in many ways, but some of these impacts will not be immediate – the application of genomics research to validating novel disease targets requires about a decade for subsequent drug development. By contrast, the application of whole genome pharmacogenetics will rapidly yield better-targeted medicines (Zimmern and Cook 2000; Roses 2003). Policy-makers accept the principle that pharmacogenetics will be a powerful impetus for rationalising therapy and reducing costs by targeting medicines to those who will benefit (Wanless 2002).

7.1
Personalised safety

Pharmaceutical companies have been proactive in considering how pharmacogenetics may change both health services delivery and pharmaceutical R&D processes: this emerging area provides a good illustration of how industry must work in partnership with academia, regulators and the health services both in identifying the conceptual issues and in the conduct of TA.

Some of the potential implications of pharmacogenetics have been acknowledged for decades, even though the genomic basis was not understood. For example, it was well known that many types of adverse drug reaction could be attributable to polymorphic gene alleles of drug-metabolising enzymes and many complex disorders (asthma, diabetes, autoimmune disease, mood disorders) exhibit large variation in patient response to standard drugs.

The primary goal for the pharmaceutical sector application of pharmacogenetics is personalised patient safety (Roses 2003). Personalised efficacy is unlikely to be achieved so precisely as safety because of gradients of therapeutic efficacy and the placebo effect; thus the population would be segmented for efficacy, rather than individualised. Commercial concerns have been raised that the application of pharmacogenetics to efficacy would fragment the market – but identifying those patients in a population who respond well to a drug should lead to fewer failures during pharmaceutical R&D and so increase the number of effective drugs appearing onto the market.

7.2
Creating supportive environment for technology development

It is important to recognise that effective assessment of the full potential of an emerging technology depends not just on evaluating the body of evidence generated but also on creating the optimum conditions for the technology to be developed and realised: the training and skill needs for the research community, the strategic framework and development of public-private consortia to share expertise and standardise tools, the education curriculum for the next generation of medical practitioners. There are various initiatives underway to share pharmacogenetic information and learning. For example, the central NIH repository of information on variation in human genes associated with variation in drug response, together with an interlinked taxonomy of central concepts and techniques is on www.pharmgkb.org/do/serve?id-home.projects.

In the UK, the Government-initiated Genetic Knowledge Parks are designed to encourage industry and academia to work together on the wider social issues for pharmacogenetics and other emerging technologies. A recent Report from the University of Cambridge Public Health Genetics Unit (now part of Cambridge Genetics Knowledge Park, www.phpc.cam.ac.uk/epg/IPP.html) calls for a strategic approach to pharmacogenetics that gives guidance to industry and encourages public sector expertise. Among the pharmacogenetics policy points raised by the Report ("My very own medicine: What must I know?) are:

- It is currently premature to devise detailed regulations as the science is still unclear;
- The establishment of a clinically-relevant evidence base for tests and test-drug combinations is a priority. Tests should be subject to validation by formal clinical appraisal;
- A balanced public policy framework for "orphan" tests and drugs (see later) must be developed;
- Public sector expertise requires strengthening (for example – further investment in clinical pharmacology);
- Current uncertainty about confidentiality safeguards could constrain research, and balanced guidance is needed (see later);
- Pharmacogenetics will add to the complexity of prescribing – "expert" systems are needed to aid the safe and effective use;
- Post-marketing surveillance (for dugs and tests) needs to be strengthened, creating a national surveillance system.

7.3
TA – ethical, social and legal issues

The first examples of the application of pharmacogenetics in clinical research and practice are appearing and it has been predicted that biochip-based testing will be standardised by 2010–2015 with the data routinely available in primary care for targeting drugs. Some researchers maintain that pharmacogenetic testing does not raise any fundamentally new social issues and that its application will be less contentious with regard, for example, to privacy and confidentiality concerns than other forms of genetic testing. Clearly, however, TA of pharmacogenetics must cover a range of ethical, social and legal issues, for example with respect to: health care delivery (skills needs, information handling, quality assurance); funding of services; equity of testing and treatment provision; public research agenda setting. Evaluation of the economic impact will also be of central importance in TA.

The ethical issues associated with pharmacogenetics are currently the subject of a detailed examination by the UK Nuffield Council on Bioethics, to which industry and other stakeholders are contributing information and advice. The contribution by the pharmacy community (Royal Pharmaceutical Society of Great Britain 2003) is of particular interest and relevance in framing the development of this technology. The Nuffield Council on Bioethics is consulting on a range of TA issues for medical practice as well as ethics (Nuffield Council on Bioethics 2002):

– Will the applications of pharmacogenetics increase inequalities in the provision of health care?
– What are the implications of finding a genetic variant that influences the response to a medicine in a particular racial or ethnic group?
– Should the primary care physician be responsible for providing a test or should tests be available to patients "over the counter"?
– New tests will require the large-scale use and storage of genetic information. What regulations will be necessary to ensure that patients' confidential information is treated appropriately and consent to its use obtained?

This latter point is important in linking the field of pharmacogenetics into the general issues for genetic data base construction and use. Several current or Accession EU Member States (for example, UK, Estonia) are developing population genetic data bases. Commission-funded research is mapping data base activity across the EU in order to identify key ethical and social issues. One issue raised recently (Pullman & Lotus 2003) suggests extending to pharmacogenomics research the principle of benefit-sharing (as initially advanced by the Human Genome Organisation) – representing the view that the genome is the common heritage of all and that requirements of distributive justice mandate benefit-sharing. However, before assumed commercial profits can be shared, it is also worth emphasising another point made by Pullman & Lotus: that there is much uncertainty about the pharmacogenomics future – both in terms of the economic costs of development and implementation of innovation and of the economics and health benefits that may accrue.

Bioscience-based companies are aware of the ethical issues arising from the application of new technologies and they commit to processes of self-regulation (Fears and Tambuyzer 1999). Companies also understand that emerging technologies such as pharmacogenetics (and others, such as gene therapy, xenotransplantation) will raise additional issues such that they need to find structures to engage in dialogue with the public-at-large (balancing views of activist groups against other public concerns). Informed dialogue must, of course, represent more than Public Relations if industry-societal exchanges are not to be polarised and dramatised. And, public consultation must occur throughout the scientific research process, not just at the point of release of a new technology.

Data from the European Commission-funded Eurobarometer survey (2002) show relatively high public support across the EU for genetic test applications considered broadly (such support is lowest in Germany, Austria and the Netherlands). Generally the support for the health care applications of biotechnology is significantly higher than support for agricultural and novel food applications.

7.4
Partnership with regulatory authorities

The criticism has been expressed (in the Cambridge Report previously cited, www.phpc.cam.ac.uk/epg/IPP.html) that pharmacogenetics knowledge is being restricted to industry "silos", with a dearth of public sector and regulatory expertise and limited access to commercial research findings. It is undeniable that companies are leading in the discovery, development and assessment of pharmacogenetics, but they also recognise the importance of educating the Regulatory Authorities about the potential of the technology in order to develop consistency in regulatory regimes and facilitate the timescale of regulatory approval (Morris 2000). Thus, the driving force in TA has not been regulatory demand but rather the industry perception that commercial value is raised by better predictions of safety: the application of pharmacogenetics is likely to change the way companies approach safety testing, for example by bringing new precision to comprehensive post-marketing surveillance programmes. Industry and the regulators now have a joint responsibility in creating new standards in both efficacy and safety (Zimmern and Cook 2000). There are significant advantages to these partnership initiatives between industry and the Regulatory Authorities – industry achieves greater freedom of action and simplification of controls; the regulators achieve access to complementary expertise and information. It is, of course, essential for such partnerships to be appropriately transparent and to be structured so that the partners learn from their relationship.

A recent initiative by the European Agency for the Evaluation of Medical Products (www.emea.eu.int/pdfs/human/pharmacogenetics/444503en.pdf) is of great significance for the EU in bringing together industry and regulators to identify the issues for pharmacogenetics TA and the institution of testing procedures. In the UK, work by the Government advisory body, the Human Genetics Commission, has also been pivotal in catalysing discussion of pharmacogenetics issues between industry, academia, ethicists and regulators (www.hgc.gov.uk/business_meetings_10september.htm). Such discussions are able to move beyond the restatement of entrenched positions to challenge key assumptions, for example, to explore the proportion of adverse events currently associated with medicines use that might be reduced by implementation of pharmacogenetic testing.

7.5
HTA policy issues in pharmacogenetics: progressing the agenda

In the Cambridge study previously cited (www.phpc.cam.ac.uk/epg/IPP.html), there was a general consensus that there are few entirely novel technical issues in the assessment of pharmacogenetics. It was found debatable whether existing health services HTA capabilities could contribute to the assessment of pharmacogenetic tests, and the field of pharmacogenetics is predicted to highlight the information gap between current licensing requirements and the needs of patients and practitioners. Pharmaceutical companies will apply the technology and TA to their own, patented, products but who will perform the pharmacogenetic assessment for generic or non-prescription medicines? There is a role for the public policy maker, public research funder and academic laboratory in identifying the scope for TA and engaging in that TA for those drugs that probably will not be characterised by pharmaceutical companies (Roses 2003). If this societal responsibility is not faced, then there is risk of creating a new underclass – those who depend on generic drugs (hence new "orphan" drugs and tests).

The UK Government recently published a White Paper on Genetics (UK Department of Health 2003), promising new funding for pharmacogenetics research on existing (marketed) medicines that will help to address this "orphan" issue. The White Paper also supports the principle of partnership – calling for greater collaboration between health services clinicians, academia and industry in performing pilot assessment studies.

7.6
Pharmacogenetics as a paradigm for TA

In summary, pharmacogenetics as a field exemplifies many key features of collaborative TA and provides a basis for work across a broader front:

- The rapidly advancing area is beginning to remove initial uncertainty about the scientific basis and rate of clinical application;
- The technology has the potential to drive the revolution in genetics products and services, improving quality of health care, and also transforming pharmaceutical R&D processes;
- Companies are leading in TA in pharmacogenetics – the sharing of information and its evaluation will be key issues for regulators and policy-makers;
- There is a critical need to identify how public and private research agendas can be aligned so that best practice in methodology is shared and impact is maximised for the health services;
- The points of similarity and difference between pharmacogenetics and other forms of genetic testing should be clarified so that the appropriate regulatory oversight is developed and the wider ethical, legal and social issues are set into context;
- Greater knowledge of personal genetic information and the interaction between genes and environment will put new demands on the public and the individual with regard to personal choice and responsibility for health status. This may accelerate the transfer of duty of care from physician to patient. There is also the issue of who is responsible for damage caused to an individual as a result of research, if society has been closely involved in decision-making during the research process (UK Cabinet Office 2003).

7.7
What is different about industry TA?

Has the information provided above from the case study of pharmaceutical industry product assessment and pharmacogenetics assessment helped to clarify what is reconcilable between the public and private sectors? Health care companies will take a practical view in their own TA, linking to specific product development and integrating with the overall assessment of product value. Prior to embarking on TA, companies will evaluate medical needs and market opportunities and will have researched tools and techniques for TA. Company TA for emerging technologies, like Product Assessment, is oriented to customer (and health services) needs and there will be organisational pressures, for example relating to timescale and costs of assessment. In some cases TA may become linked with processes for business risk assessment, corporate social responsibility and defence of corporate reputation. To serve these goals, some of the operational characteristics of the company TA process may differ from those undertaken outside industry (see Introduction) but companies appreciate the central importance of developing stakeholder partnerships in order to address concerns about the perceived lack of independence and neutrality (criteria that would be expected to characterise TA in the public sector), to optimise methodology and the likelihood of its TA being accepted by the other stakeholders.

Depending on their requirements for multi-disciplinary expertise and evidence, companies will vary in their desire to embed TA within their processes or refer to external experts: that is, companies will choose between (i) conducting TA "in-house" and then sharing the outputs and (ii) sharing the TA process itself. Apart from operational considerations, such as whether the competencies are available in-house, companies will select a preferred TA strategy according to their views on public sector TA. Some companies, for example will have concerns about the twin public sector dangers of institutionalising TA and over-prescribing the criteria for TA. It is the thesis of this paper that, in pharmaceutical sector TA, differences between industry and the public sector are reconcilable. Industry has TA interests in common with the regulators and health service providers in creating well-informed customers who have confidence in the quality of the deliverables.

8
Recommendations for action

In analysing the case study of the UK pharmaceutical industry, it is concluded that there are, indeed, opportunities for public-private sector formal collaboration in TA and for sharing the outcomes from TA performed in the individual sectors, in order to serve the twin goals of public policy and innovation. TA in the field of health care might be regarded as a potential example of best practice. It is important, at the same time, to respond to the criticism that the TA community (both public and private sectors) is still dominated by the linear model of innovation – the unidirectional flow of information from researcher to policy-maker. While there has to be a good linear dissemination of the outputs of technology and innovation, it is also imperative to involve research users at all stages of R&D procurement and knowledge exchange (Editorial 2003).

Therefore, to promote European competitiveness, it is no longer enough merely to increase the supply of ideas by investing in R&D and facilitating technology transfer. What, then, should now be attempted in disseminating and interrogating the assumptions from this and other industry TA case studies? The European Commission updated policy strategy argues for a multidimensional and horizontal ("third generation") approach in which innovation will become central to all policy areas (DG Enterprise 2003). As a result of the TAMI project, a coordinating Platform can now be proposed to support analysis of the role of TA in innovation, and Commission funding should be sought.

What are the opportunities? A TA Platform would be relevant to many proposed European research activities – the current Commission-funded Framework Programme 6 calls for assistance and support for the coherent development of research and innovation policies. Strategic work is solicited, for example (http://ted.publications.eu.int/static/doccur/en/en/114412-2003.htm) in: R&D econometrics, technology innovation, research-industry links, industry R&D management and productivity, interaction between regulations/standards/research/innovation, Foresight methodologies and practice. In particular, strategic support actions across the thematic priority "Life sciences, genomics and biotechnology for health" will cover the facilitation of assessment, experience-sharing at the public-private junction, debate on technology options, social scrutiny and dialogue (http://fp6.cordis.lo/fp6/call_details.cfm?CALL_ID=88).

The recent EU Greek Presidency also outlined a series of guidelines to further the implementation of a European Research and Innovation Area (ftp://ftp.cordis.lu/pub/greece/docs/eria_gsrt_2003_en.pdf), assumed necessary if the Lisbon and Barcelona Summit goals for innovation and competitiveness (R&D spending to reach 3% of GDP) are to be attained. In order to embed TA within the multiple Commission-funded activities on innovation (particularly in the health and life sciences), and help to create an encouraging environment in which industry will want to increase its own R&D investment, it is recommended that TAMI now develop further the concept of an open-membership, Commission-funded TA Platform as a means to:

- Share best practice in TA;
- Refine the taxonomy of TA criteria and instruments;
- Support horizon-scanning (for example, emerging technologies);
- Identify additional mechanisms for collaborative work between stakeholder constituencies;
- Contribute to improving the environment for R&D across the EU (and incentivising companies to increase R&D investment).

It would not be the purpose of the Platform to enforce consensus in TA practice or to ignore the scepticism expressed (Hennen 1999) that industry can ever fully take into account negative side effects of the innovation process. Rather, the goal is to identify a common starting point and to debate tractable, transparent, decision-oriented methodologies that allow the stakeholders to realise their own agendas. This combination of the standardisation of tools with a commitment to pluralism in processes has worked well in some of the national Foresight exercises, providing overall credibility where, necessarily, there may be a divergence of views on outcomes. Constitution of a proactive Platform now would be timely in preparing for Enlargement and might also be considered as a first step in a more comprehensive and integrative approach to global S&T policy cooperation (Stein 2002). While the engagement of multiple industry sectors with a Platform will be a challenge, it may yet prove a greater challenge to integrate the interests of society-at-large without risking domination by special Interest NGOs.

References

Becks H, Gelbke HP (2001) Die Ökoeffizienz-Analyse nach BASF. In: TA-Datenbank-Nachrichten, No 2, 10. Jahrgang – June 2001, p 34

Bouchier A (2003) Industry ponders reimbursement crisis. Nature Biotechnology, 21, 347–348

Brochler S, Simonis G, Sunderman K (eds) (1999) Handbuch Technikfolgen-Abschätzung, Sigma, Berlin

Buxton M, Hanney SJ (1998) Evaluating the NHS research and development programme: will the programme give value for money? Royal Society of Medicine, 91 (suppl 35), 2–6

Cabinet Office (2003) Biosciences: challenges and opportunities for Government, Strategic Futures on www.strategy.gov.uk

Canning D (2003) New technology in health care. Science in Parliament, 60, 8–9

Chapman S, Reeve E, Rajaratnam G, Neary R (2003) Setting up an outcomes guarantee for pharmaceuticals: new approach to risk sharing in primary care. British Medical Journal, 326, 707–709

Comroe JHJr, Dripps RD (1976) Scientific basis for the support of biomedical science. Science 192, 105–111

Dargie C (2000) Policy Futures for UK Health. The Nuffield Trust

Department of Health (2003) Our inheritance, our future. Realising the potential of genetics in the NHS

Department of Trade and Industry (2002) The 2002 R&D Scoreboard

DG Enterprise (2003) Innovation Tomorrow, Innovation Paper 28, www.cordis.lu/innovation-policy/studies/gen_study7.htm. Editorial, Admitting the evidence, Innovation Policy Review 2003 4, 1–4

ESRC (2001) Innovative Health Technologies Programme 2001 on www.york.ac.uk/res/iht

Fears R, Poste G (1999) Building research resources in human population genetics: the potential of the British National Health Service. Science 284, 267–8

Fears R, Tambuyzer E (1999) Core ethical values for European bioindustries. Nature Biotechnology 17, 114–115

Fears R, Roberts D, Poste G (2000) Rational or rationed medicine? The promise of genetics for improved clinical practice. British Medical J. 320, 933–935

G10 Medicines Group, High Level Group on innovation and provision of medicines in the European Union. Recommendations for action, May 2002 on http://europa.eu.int/comm/health/ph/key_doc/key08_en.pdf

Gelijns AC, Rosenberg N, Moskowitz AJ (1998) Capturing the unexpected benefits of medical research. New Engl. J. Med. 339, 693–698

Grunwald A (2002) Technikfolgenabschätzung – eine Einführung. Sigma, Berlin

Hennen L (1999) Technology Assessment – does it always come "too late"? In: TAB-Brief, No 17

Lasker Charitable Trust, Exceptional Returns – The Economic Value of America's Investment in Medical Research, 2000 on www.fundingfirst.org

Martin S (2003) The evaluation of strategic research partnerships. Technology Analysis and Strategic Management 15, 159–176

Medical Technologies Group, Enhancing patient access to new medical technologies. Science in Parliament 2003 60, 20–21

Miller DH (2003) Commentary: Evaluating disease modifying treatments in multiple sckerosis. British Medical J. 326, 525

Minx E (2001) In: TA-Datenbank-Nachrichten, No 2, 10. Jahrgang, p 39

Morris N (2001) The changing landscape of regulatory control of biological medicines. Technology Analysis & Strategic Management 13, 246–263

NIH, NIH GPRA Research Programmes Outcome FY 1999 Assessment Material on www1.od.nih.gov/gpra accessed on 3 December 2002

Nuffield Council on Bioethics, Consultation on pharmacogenetics: ethical issues, 2002 on www.nuffieldbioethics.org/pharmacogenetics/public.asp

OECD, The OECD Health Project relating to new and emerging health related technologies on www.oecd.org accessed 24 March 2003

Pardes H, Manton KG, Lander ES, Tolley D, Ullian AD, Palmer H (1999) Effects of medical research on health care and the economy. Science 283, 36–37

Pharmaceutical Industry Competitiveness Task Force, Final Report March 2001, Department of Health and Association of British Pharmaceutical Industry websites

Postnote 168 MS treatments and NICE, Parliamentary Office of Science and Technology 2002

Pullman D, Lotus A (2003) Clinical trials, genetic add-ons, and the question of benefit-sharing. Lancet 362, 242–244

Richards T (1999) The genomic challenge. British Medical J. 318, 341–342

Ropohl G (1999) Innovative Technikbewertung. In: Brochler S, Simonis G, Sundermann K (eds) Handbuch Technikfolgen-Abschätzung. Sigma, Berlin

Roses A (2003) Pharmacogenetics: personalised safety and segmented efficacy on www.acmed-sci.ac.uk/forum_roses.htm and related articles cited therein

Royal College of Physicians (2000) The prescribing of costly medicines

Royal Pharmaceutical Society of Great Britain, Nuffield Council on Bioethics. Consultation on pharmacogenetics: ethical issues on www.rpsgb.org.uk/pdfs/pharmacogene.pdf accessed on 17 February 2003

Salmankaita JP, Salo A (2002) Rationales for government intervention in the commercialization of new technologies. Technology Assessment & Strategic Management 14, 183–200

Salo A, Kuusi O (2001) Developments in parliamentary technology assessment in Finland. Science & Public Policy 28, 453–464

Stein JA (2002) Science, technology and European foreign policy: European integration, global interaction. Science & Public Policy 29, 463–477

Steinmuller K, Tacke K, Tschiedel R (1999) Innovationsorientierte Technikfolgenabschatzung. In: Brochler S, Simonis G, Sundermann K (eds) Handbuch Technikfolgennabschätzung, Band 1. Sigma, Berlin 129–145

Vaccani P, Bax R, Watson P (1997) Measuring outcomes from R&D in healthcare. In: Anderson J, Fears R, Taylor B (eds) Managing technology for competitive advantage. Cartermill, London 379–399

Wanless D (2001) Securing our future health: taking a long-term view on www.hm-treasury.gov.uk/Consultations_Legisl.../consult_wanless_index.cf

Weber J, Schaffer U, Hoffmann D, Kehrmann T (1999) Technology Assessment – Eine Managementperspektive – Bestandsaufnahme, Analyse, Handlungsempfehlungen. Gabler, Wiesbaden

Zimmern R, Cook C (2000) Genetics and Health. The Nuffield Trust

Culturally-Based Framing Factors that Influence Technology Assessment

Tomasz Szapiro

1
Introduction

Cultural differences in our era of globalization seem to be an unavoidable background of social processes. According to Britannica *culture is behavior peculiar to Homo sapiens, together with material objects used as an integral part of this behavior. Thus, culture includes language, ideas, beliefs, customs, codes, institutions, tools, techniques, works of art, rituals, and ceremonies, among other elements.* In this analysis culture is defined in a more narrow sense to describe behaviors related with technology assessment with an implicit focus on compromising societal projects[17].

Actors who are involved in these processes frame problem situations including the cultural dimension. The *frame of reference*[18] affects the results of processes. Moreover, *framing* meant as an act, process, or manner of constructing anything is also considered as a tool in solving process-related problems and dilemmas (e.g. Kahneman and Tversky (1991) papers of prospect theory and framing; Simon's (1976) or March's (1988) books on organisational contexts of decision making).

Different ways of framing situations arise as a result of the different (sub)cultures of the parties involved in projects. Professional management of a project requires consideration of the *framework* – a collection of shared values and beliefs held by members of a group – as an important part in the planning and operations phases of project (Phatak 1995). The postulate of building a framework is a joint task of the involved parties. Frequently frameworks are pre-assumed (everybody has some idea of the vision of the others), inherited (frequently actors are invited to projects having mature and hard frames) but there are also situations when actors are in a position to influence the framing. This includes individual evaluation of cultural differences followed by the group making compromises to reach an agreed framework.

To make the terminology clear let us define *outcome* as an identifiable result of the project and *objective* as the goal or aim of the program's staff, or at least of someone with interest in the program. Some outcomes may occur unintentionally. Following the literature on impacts, the term *problem* is used for a condition that

---

[17] In the chapter the precise meaning of the term culture is national culture or as institutional (organisational) culture, see also Deal and Kennedy (1983) and Schein (1990). It assumed that the context decides which meaning is relevant if clear definition is not given.

[18] I.e. a structure of concepts, values, customs, or views by means of which an individual or group perceives or evaluates data, communicates ideas, and regulates behavior.

from some perspective would be considered both unsatisfactory without the intervention of the program in question and satisfactory or at least more acceptable given that intervention (see Fischoff et al. 1991).

Technically the word *program* means a perfectly valid perspective and it is defined by an outcome (target). This meaning will be extended in the document to a large number of outcomes in the form of concerns both of program personnel and of interested outsiders. Intended and unintended impacts are considered. They may not be seen in connection with the program until new times or new events suddenly make them salient and the connection is visible.

Future events are troublesome. This element of analysis can never be observed and can never be known for certain. Its importance in assessing the impact of a program and its inaccessibility makes this the central point of all impact-analysis designs. It is also a major point of reservations about the validity of evaluative conclusions. The term *counterfactual* addresses the future dimension in detailed analyses.

The following five problems[19] or functions connected with the effectiveness of multiple-outcome programs are recalled in the literature: finding, limiting, assessing impact, common scaling, and weighting. *Finding* aims at identifying (frequently a large number of) impacts. This raises the problem about how to figure out which outcome dimensions are important (to identify objectives, constraints and their importance, side effects). *Limiting* the number of huge potential impacts calls for evaluating impacts of all of outcomes, and deciding which seem to be unrealistic, raising the problem of how to decide which to consider and which to neglect. *Assessing impact* requires discussion of the methods, in particular how to estimate the impact of the program (future activities) on each outcome. Need for *common-scaling* results from questions about the effectiveness of the project with various outcomes combined in some way. This raises the problem of how to put all aspects together to raise the effectiveness of the whole. *Weighting* is needed to integrate the varying importance of the impact scores for combined outcomes. However, assessing the importance of each outcome relative to the others is problematic.

Project TAMI attempts to explain the process of setting a value on technology for the purpose of evaluating the impact of this process. In most countries central government agencies carry out the assessing. The analysis presented in previous chapters, supplemented by the examples, has important practical conclusions. The applicability of these findings should be reviewed from point of view of the known psychological and sociological mechanisms that influence decision making processes. This paper attempts to answer this call. Some important concepts related to assessment are discussed from three perspectives. Since people who tackle real problems can be treated as individuals, but also viewed as members of communities, viewpoints of an individual, of an organisation and of a nation are considered. Firstly, factors influencing problem and impact identification are reviewed with a focus on failures in problem recognition and so called outcome definition. Next, organisational culture and its functions are recalled. Finally, the problem of assessing cul-

---

[19] The terminology is recalled for the use in this chapter – there are also other meanings of used terms.

tural differences between countries is considered within the Kluckhohn-Strodtbeck framework and the Hofstede framework. The body of knowledge briefed in the first part of chapter allows suggestion of a way to include cultural factors in TA roles investigation and operation.

2
Factors influencing problem and impact identification

2.1
Failures in problem recognition

TAMI proposes a common framework to enable the understanding of the relationship between method and impacts. This framework assumes the structure (presented in the chapter on methods) of technology assessment. However as impact can only be achieved by TA-projects, which implement a project design, it is important to understand the sources of the differences between final technological solutions and targets used in their presentations, verification and the initial problem statements (see also Decker and Grunwald 2001). The process involves project design, choice of adequate methods from a "methods toolbox" and establishment of criteria. Criteria are organized in clusters (scientific, interactive and communication) according to the relevant interdisciplinary contexts, processes and social environments. The project design involves the crucial steps of the TA-structure explained earlier, please see the figure below.

In order to avoid *false identification or deformation of the problem* it is essential to supplement the project design for a particular outcome with questions about the problem and a range of counterfactual identifications. Reports point out that the outcome may represent only one possible way of tackling a problem – not necessar-

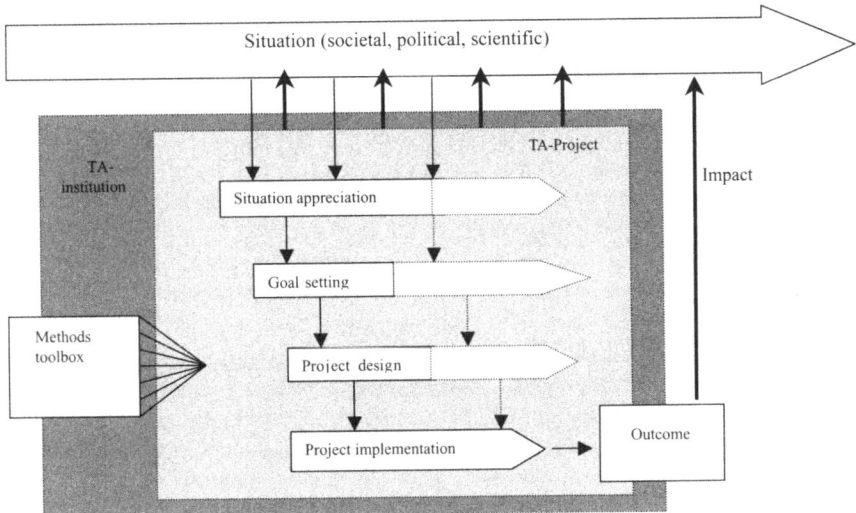

Fig. 1: From Method to Impact

ily adequate – and outcome definition may distract attention from the problem itself in its larger scope. Frequently, analysts are too focused on technicalities- numbers etc. Focus on problem may be also achieved through inclusiveness of the population that is the program's target. This idea (of the so called *coverage*) seems to be evident when the problem of homelessness is considered as an example in it complexity. Focusing on the desired outcome is an easy and commonly used way to orient the impact analysis but it is highly unlikely give positive effects.

Also a fixation on outcomes frequently leads to a *failure to recognize multiple concerns* (see e.g. Berleman and Steinburn 1967). What may seem to be one outcome can actually represents several possible problems demanding alternative program activities. The so-called *juvenile delinquency* is an example of where measurements believed to capture this concept are widely discussed but different problem dimensions are obviously relevant but are not considered. This kind of confusion is not so likely to happen if attention is focused on the problem itself and not on routines. Failure here can lead to ineffectiveness and inefficiency of undertakings and their evaluators.

Another source of failure is the incorrect *structuring of problems*, see Keeney and Raiffa (1976). There may be a tendency to state and measure the wrong (incidental) outcomes. This leads to confusion, poor program planning, and a waste of evaluation efforts. For example, let us consider the case of elderly people who cannot work easily. In this case the program focused on training and education but discarded the expensive problem of improving the health of these people. The policy failed to recognize this dimension. Neglect of thinking about the role of the counterfactual and the aims of impact analysis is generally a problem. Attention to this problem is a way to assure that counterfactual will influence program evaluation. For example, given the demand for 1000 new consumers, having 750 may be seen as 75 percent effective. But that statement neglects the issue of the number of potential of customers: new ones unequally create firm profit.

2.2
Failures in outcome line definition

Thinking in terms of outcomes, outcome line representing a target is unquestionably included in any impact analysis. However, almost anything might be a legitimate outcome and this can perturbs appropriate problem definition see Earley et al. (1990). Self-questioning definition of a problem increases the likelihood of efficient impacts later. A typical gap relates to the trap of evaluating a program only for its effects upon those it actually serves. Those effects do represent an outcome. When one specifies what would be inadequate without the program, however, the necessity for an outcome of broader scope will often become clear.

Programs may have multiple outcomes of interest. Multiple outcomes may be consistent with one another, but they may also be conflicting. Conflict among outcomes is one of the primary reasons for the weighting and common-scaling functions in multiple-outcome evaluation. There are evaluation techniques that confront the same set of issues – in cost-benefit analysis, see e.g. Gramlich (1981), and in multiattribute-utility methodology – MAUT, see e.g. Edwards and Newman (1982) and also Sherrill (1984).

Cost benefit analysis. Advantages of cost-benefit analyses are evident: one gets numerical indicators, which may be used, for recommending a decision rule. The *common scaling* and *weighting* are a natural tool in cost-benefit analysis (use of percentages is a typical practice, but other common scales are equally acceptable). This obliges consideration of the dangers of such analyses. For example, frequently certain benefits and costs are considered in research as result of a specific "imitation strategy". Following program conventions some benefits and costs reflect a specific background. Also a "learning curve" exists in this setting, resulting in transfer of this method from one program to another. The literature charges analysts with a responsibility to ponder more than calculative gains and losses but also indirect or distant implications. Typical examples are pollution, health, safety, waste of time, secondary market impacts, non-monetary satisfaction from education, impacts on marital and family ties etc. They should take into account traps of cost-benefit analyses related to completeness, data accessibility and correctness, double counting, or generally: dependence of characteristics considered. Cost-benefit analysis depends on assessment of *impact*. This task is difficult and complex and calls for expertise and creativity. Unfortunately, often there are benefits and costs dealt with in analyses which are not convertible to monetary values (health, cognitive development, awareness of political issues, divorce, environmental aesthetics, extinction of animal and plant species, and so forth (see also Janis and Mann, 1977).

The *multiattribute decision making* approach leads to decisions concerned with recommendations of action, given the results of an evaluation (not theory)[20]. MAUT aims at deciding which alternative is to be recommended. MAUT considers a large number of outcomes and permits their aggregation into one scalar indicator of impact. In MAUT, individuals interested in outcomes are called *stakeholders*. The stakeholders know what the relevant impacts are or should be so in practice whatever list they compose may be considered complete. There is no possibility of an incorrect evaluative conclusion that is due to the omission of impacts by stakeholders. By assumption, the interchange of ideas will help the stakeholders to arrive at a list of outcomes. MAUT uses a common-scaling and weighting system based on scalarization of all impacts and thus it is deeply subjective[21]. The participants in a decision process make up their own minds individually about the action indicated and arrive at a collective resolution by bargaining and politics instead of rational persuasion or demonstration.

Impact analysis is not a series of independent sections, but rather a map representing a logical whole. Frequently specific procedures are used to carry out appropriate evaluations. They usually include:

- *Comprehensiveness in the presentation of outcomes.*
- *Correct choice of the outcomes that are to be submitted to actual research.*
- *Duality of selection and execution of research.*

---

[20] MAUT is a tool for organisational decision-making. It is derived from the goals of members of the organisation having a stake in the decision because of their official responsibility.

[21] One argues here that using averages decreases the disadvantages of subjectivity in common scaling and weighting. However, in collective decisions (in public administration and public policy) averages of values can rarely be satisfactory and acceptable.

Impact analysis does not determine all elements of policy processes (like politics, bargaining, and trade-offs with other policies, personal considerations). Analysis possibly should provide good information – in the form of explained recommendations. Valuable analysis reveals the implications of policy alternatives.

These observations may be employed to build routines, which feature the relationship of the problem to the outcome line, and give technical devices for avoiding some types of difficulties related with evaluation of the right questions. Let us observe that difficulties that correspond directly to the fact that requested outcomes of interest and factual outcomes of interest need not be a conscious objective.

3
Culture

The projects under considerations are led by organisations. The project result includes individual actions. Culture – norms, habits, attitudes although hard to define precisely – plays an enormously important role in decision taking. Culture limited to an organisation constrains and emphasizes social influences. Most large organisations have a dominant culture and numerous sets of subcultures. A strong culture can act as a substitute for formalization regulating employee behavior and creating predictability, orderliness, and consistency. Formalization and culture can be viewed as two different solutions of the ordering problem. As discussed in the chapter on POTA – Parliamentary Technology Assessment – the institutional context of technology assessment activities, including the interaction between politicians' demands and analysts' strategies, are key factors in the general explanation of impacts (see also Liakopoulos 2001). Also, existence of contradictions between the institutional arrangements and the relative influence of the technological assessment was pointed out. The degree of autonomy and exclusivity of the assessment adds to a better understanding of diversified impacts. However, perception of autonomy or exclusivity is culturally dependent and framing actors' attitudes is an important part of context perception and definition.

3.1
Organisational culture and its functions

Organisational culture refers to shared meaning that distinguishes the organisation from other organisations. This shared meaning involves:

- *Individual initiative:* the degree of responsibility, freedom, and independence that individuals have;
- *Risk tolerance:* the degree to which employees are encouraged to be aggressive, innovative, and risk seeking;
- *Direction:* the degree to which the organisation creates clear objectives and performance expectations;
- *Integration:* the degree to which units in the organisation are encouraged to operate in a coordinated manner;

- *Management support:* the degree to which managers provide clear communication, assistance, and support to their subordinates;
- *Control:* the number of rules and regulations, and the amount of direct supervision that is used to oversee and control employee behavior;
- *Identity:* the degree to which members identify with the organisation as a whole rather than with their particular work group or field of professional expertise;
- *Reward system:* the degree to which reward allocations (that is, salary increases, promotions) are based on employee performance criteria in contrast to seniority, favoritism;
- *Conflict tolerance:* the degree to which employees are encouraged to air conflicts and criticisms openly;
- *Communication patterns:* the degree to which organisational communications are restricted to the formal hierarchy of authority.

Appraising using these ten characteristics forms a composite picture of an organisation's culture. Individuals with different backgrounds at different levels in the organisation will tend to describe the organisation's culture in similar terms. Culture has a defining role – it creates distinctions between one organisation and other (see Adler 1991). Culture conveys a sense of identity for members of an organisation and facilitates the generation of commitment to something larger than one's individual self-interest. Culture enhances the stability of a social system thus helping to hold the organisation together. Culture serves as a sense-making and control mechanism that guides and shapes the attitudes and behavior of employees (see e.g. Knotts 1989).

Culture can be viewed as a liability. One should not ignore the potentially dysfunctional aspects of culture, especially a strong one, such as an organisation's effectiveness. Culture is a liability where the shared values are not in agreement with those that will further the organisation's effectiveness. When the environment is undergoing rapid change, the organisation's entrenched culture may no longer be appropriate. So consistency of behaviour is an asset to an organisation when it faces a stable environment. It may, however, burden the organisation and make it difficult to respond to changes in the environment.

Organisational culture – current customs, traditions, and general way of doing things – are largely due to what it has done before and the degree of success it had with those endeavors. The founders of an organisation have a major impact in establishing the early culture (Meek 1988). Therefore the culture results from the interaction between the founders' assumptions and learned experiences. Once a culture is in place, there are practices within the organisation that act to maintain it by giving employees a set of similar experiences. Cultural aspects can be supportive in impact evaluation. Especially, where a dominant culture exists, it can become very resistant to change.

3.2
Assessment of cultural differences between countries

The different social practices reflect different cultures and result in different behaviors. Evidence suggests that children are preprogrammed in the ways of its culture by the time they are adults. They understand how things are done and can work comfortably within their country's unwritten norms, but they cannot explain their culture to someone else. Most people are unaware of just how their culture has shaped them. However there are findings that explain how cultures vary. For example, in the context of international negotiations some factors are identified that influence interactions and thus can limit or constrain international organisations. These factors are: political and legal pluralism; international economic factors; foreign governments and bureaucracies; and instability and ideology. These, however, are descriptive categories. In order to explain mechanisms one needs to consider different frameworks.

3.3
The Kluckhohn-Strodtbeck framework

One of the most widely referenced approaches for analyzing variations among cultures is the Kluckhohn-Strodtbeck framework (see Kluckhohn, 1954 and Kluckhohn and Strodtbeck 1961). It identifies six basic cultural dimensions: relationship to the environment; time orientation; nature of people; activity orientation; focus of responsibility; and conception of space. People and countries evaluated in this way appear to differ strongly.

Relationship to the Environment. Attitude to dominate their environment influences organisational practices. In a conquering society, goal setting is not popular and expected penalties for failure are high. In a harmonious society deviations from goals are accepted.

Time Orientation. Time in the West is a scarce resource. This results in the short-term orientation in performance appraisals. Past oriented cultures follow traditions and preserve old practices. Culture's orientation is reflected in the importance of deadlines or the length of job assignments.

Nature of People. Perception of good or evil is culture dependent. E.g. culture-oriented view on the nature of people influences the dominant leadership style of managers (autocratic vs. participatory).

Activity Orientation. Cultures can emphasize *action, being* or *controlling.* They appreciate hard work rewarded recognition for accomplishments, emotions and rationality, respectively.

Focus of Responsibility. Individual responsibility takes care of himself or herself. *Group responsibility means* sharing values and emphasis on group loyalty. *Hierarchical responsibility* implies decision-making style, communication patterns, rewards, and selection practices.

Conception of Space. Some cultures are very open and conduct projects in *public.* Others – *privately.* Mixed space orientation has implications for organisational concerns.

3.4
The Hofstede framework

Hofstede surveyed 160,000 employees in sixty countries who all worked for a single multinational corporation in order to reliably attribute variations that he found between countries to national culture. The data confirmed that national culture has a major impact on employees' work-related values and attitudes Hofstede identifies four dimensions of national culture: (1) individualism versus collectivism; (2) power distance; (3) uncertainty avoidance; and (4) masculinity versus femininity (see e.g. Hofstede et al. 1989).

Individualism versus Collectivism. Individualism is related to a personal focus on one's own interests. On the contrary, a tight social framework in which people expect others to look after them when they are in trouble characterizes collectivism. They return this protection with loyalty to the group. It appears that rich countries are very individualistic. Poor countries are very collectivist.

Power Distance. Physical and intellectual abilities create differences in wealth and power. The *power distance* as a measure of the extent to which a society accepts the fact that power of institutions and organisations is distributed unequally. A high power distance society accepts wide differences.

Uncertainty Avoidance. Response to uncertainty is culture dependent. In some societies people feel less comfortable with risks – which is described as high *uncertainty avoidance* of society. High uncertainty avoidance goes with increased level of anxiety among its people and defense mechanisms to provide security and reduce risk. Organisations follow formal rules and behaviors.

Masculinity versus Femininity. In theories of culture masculinity of a society describes a high degree of assertiveness and the acquisition of money and material things. Femininity of the society describes an emphasis on relationships, concern for others, and the overall quality of life.

An understanding of differences between cultures in interactions of people with different cultural backgrounds[22]. Selected empirical data ere presented in Table 1. The research shows that there are several areas of thought processes where culture intervenes.

Problem definition. The definition of a situation as a problematic one can differ greatly across cultures.

Building teams. The criteria used to select team members vary across cultures. The criteria for a group can include knowledge of the relevant subject matter, seniority, family connections, gender, age, experience, and status. Different cultures weight these criteria differently, leading to varying expectations about what is appropriate.

---

[22] As examples of cultural differences and similarities among nations serve United States, Australia, Great Britain, Canada (high individualism); Colombia, Venezuela, Pakistan, Peru (high collectivism); Philippines, Mexico, Venezuela, Yugoslavia (high power distance), Austria, Israel, Denmark, New Zealand (low power distance), Greece, Portugal Belgium Japan (high uncertainty avoidance); Singapore, Denmark, Sweden, Hong Kong (low uncertainty avoidance); Japan, Austria, Venezuela, Italy (high masculinity); Sweden, Norway, Yugoslavia, Denmark (high femininity), see also Grey and Thone (1990).

Table 1 Ranking of Countries/Cultures influence

	Rank Order On			
Country	**Power Distance**	**Individualism/ Collectivism**	**Masculinity/ Femininity**	**Uncertainty Avoidance**
Arab Countries	7	26	23	27
Belgium	20	8	22	5
Denmark	51	9	50	51
East Africa	21	33	39	36
France	15	10	35	10
Germany	42	15	9	29
Great Britain	42	3	9	47
India	10	21	20	45
Iran	29	24	35	31
Italy	34	7	4	23
Netherlands	40	4	51	35
Portugal	24	33	45	2
South Africa	35	16	13	39
Spain	31	20	37	10
United States	38	1	18	43

Protocols. Cultures differ in the degree to which protocol, or the formality of relationships is important. The use of first names and ignoring titles generally reveals
less formal attitudes.

Communication. Cultures influence the way that people communicate, both ver-
bally and nonverbally. There are also differences in body language across cultures;
the same behavior may be highly insulting in one culture and completely innocuous
in another.

Time. Culture has a great effect on defining what time means and how it affects
function. This is visible in actions (appearing for meetings on time) and evaluations
(of wasting the time).

Nature of Agreements. Culture affects concluding agreements. Agreements can be
logically justified (e.g. cost analyses), but also may be based on intangible values
(family or political connections). Agreements may have different meanings.

In the chapter on POTAs' impact, the role of an organisation's capacity to build
coalitions and to access decision-makers was shown. Both capacities deeply depend
on communication and building teams dimensions from the Hofstede Framework.

As we see from the data shown in the tables in this section, these dimensions vary from culture to culture. This explains why different institutional settings, together with the rules of the game and incentives structure, are factors significantly influencing impact. The complex structure of the concurrent influence of coalition building and communication, taking into account also political centralisation in the legislature, party discipline and political mandate, can be analysed in the Kluckhohn-Strodtbeck framework as affected by cultural dimensions of nature of people and activity orientation.

Among hypothesized impact determinants of POTAs – perceptions of power, risk, quality, neutrality and external communication – these depend strongly on framing cultural aspects. Indirectly, the density of the external networks and consequently the autonomy are also culturally dependent. Capturing of politicians' attention and thus information processing depends on how cultural roots influences perceptions and attitudes such as trustful behaviours (see Bazerman 1991).

Cultural dependencies related to social aspects are illustrated by the example of role "d" – "Policy objectives explored". The example deals with the replaceability of human beings. The problem of the overall performance of the robot appears crucial for project development. Steps of the TA-structure involve the cost-benefit-analysis, legal aspects and ethical aspects. Description of both frameworks of cultural factors analyses from Section 3 provides man premises for the occurrence of different problem framings. Differences result from different cultural orientations and may lead to contradictions between the institutional arrangements and the relative influence of the technological assessment.

Another similar illustration (for policy aspects) is provided in example of role "p" – "New orientation in policies established". The example deals with research policy for sustainable development. The Committee for Education, Research and Technology Assessment of the German Parliament investigated the influence of the demands that the principles of sustainable development placed upon research and technology policy. Elaboration of criteria for sustainability of research and technology development provided a framework for a reorientation of R&D policy. Again, a list of cultural dimensions influencing this framing can be easily formed.

Differences among roles can be explained based on data from Section 3. Here one can argue that the stakeholders in the debate represent different organisational cultures and frame problems accordingly with resultant attitudes. Globalization of economic and social processes leading to integration of different national backgrounds enforces this statement.

In addition to these variables that are external to TA organisations, there are other internal factors, related to these organisations' information production process, which have to do with organisational capacities (budgetary and human resources) that must also be borne in mind when analysing the impact.

4
TA roles descriptions and culture influences

Following the definition discussed in the chapter 'Method of Technology Assessment' – "...*Technology Assessments (TA) is a scientific and communicative contribution to the solving of technology related social problems...*". The matrix of typology of roles of TA assists in analyzing and understanding structural factors influencing impacts and forming technology assessment. Findings in the area of cognitive and social psychology, organisational behavior and international management (reviewed in previous sections) provide a body of knowledge which can be used to understand how framing of cultural differences may influence the technology assessment. Thus it is possible to enrich TA strategies with a cultural dimension and this will decrease the occurrence of bias related to different perceptions. The role setting of TA processes enables characterization of TA on a two-dimensional map with the coordinates "issue" and "impact". Issue aspects refer to science and technology, society and policy. Knowledge, attitudes and actions are considered to be impact aspects. Each pair of "issue – impact" aspects is in a sketchy way represented by the field in the Table 2 (please see next pages), described in detail in the previous chapters and supplemented with examples.

Table 2 simplifies the purpose-structural analysis of technology impact. Firstly, each of the fields represents a set of very different relationships that build impacts[23] . These relationships can be interdependent – for example the rise of the issue of legalisation of homosexual relationships or genetic experiments leads to newly-formed attitudes and existing attitudes influencing the perceived importance of the issue and the impact of the final result). Secondly, in order to clarify the methodology, cultural background is neglected in this analysis, while – as it was argued above – it can significantly change the impact of a project. In order to fill this gap, a scheme for investigating the cultural dimension of the impact analysis is suggested and discussed in the next two sections.

4.1
The approach

The framework for program evaluation that is reported in the literature has three major phases. Firstly, the theoretical elements are considered: impact analysis consists largely on observations about these elements and relates them to one another. The primary elements are the problem, the activities and the outcome of interest. A sound working knowledge of these concepts and the differences among them, as well as their respective functions both in the conduct and evaluation of the program are critical for the development of good program evaluation. Considerable attention should be devoted to elaborating and clarifying all elements and their offshoots to foster a useful intuitive grasp of the impact analysis scheme.

Secondly, there must be some means for determining whether the theory is correct. Whatever means are used can be called a design. There are advantages and disadvantages of different impact-analysis designs using different means.

Thirdly, a way for quantifying a program's effectiveness, if exists, is crucial. A regression coefficient, a difference between two means or two proportions, and tests of the statistical significance are examples of quantifying measures. The *effec-*

---

[23] For the sake of completeness let us recall that fields in the Table 2 were investigated according to the scheme driven by the following questions: What is the role about? What are needs for this role/impact? What are examples of such impact/role? What are external factors and internal strategies.

tiveness ratio and the *adequacy ratio* are other examples. In quantifying a program's effectiveness one must determine whether the measure refers to the population actually treated or the population at risk.

The framework presented in the chapter on TA-methods is different. However, similar to the three-phase framework, it admits involvement in analyses of cultural influences and resultant bias. In order to include the influence of culture on forming impacts, we need to assume that it is possible to identify factors responsible for cultural dependencies. The culturally rooted factors that influence impact can be classified (mapped) using two dimensions: "tangibility" vs. "directness". Accordingly, the first dimension splits factors into tangibles and intangibles. The second one takes into account the spontaneous nature of influence of a factor versus the possibility of control of this influence.

The number of relevant and current regulations in an institution can be considered as an example of tangible cultural factors. Depending on the type of attitude toward responsibility (focus of responsibility in the Kluckhohn-Strodtbeck framework) it influences important assessment technology aspects: decision-making style, communication patterns, and selection practices.

Intangible cultural factors can be exemplified using one of the Hofstede framework dimensions – masculinity versus femininity of a community. The degree of assertiveness, concern for others, the overall quality of life, are not tangible but do influence assessment procedures.

Controllable cultural factors are related, for example, with time orientation from the Kluckhohn-Strodtbeck Framework. Let us consider the use of deadlines to control the dynamics of the assessment process as an example of using this factor. The effect of a negotiator's sex in international negotiation (in negotiation with parties from eastern countries women have serious difficulties in playing their professional roles) can be recalled as another example. This factor can be deemed as controllable (through replacements of roles, tasks or members in negotiation team). This example serves also to illustrate intangible influence.

Spontaneous cultural factors are related to the nature of people identified by the Kluckhohn-Strodtbeck Framework. Perception of good or evil is culturally dependent and cannot be controlled.

Table 2 visualises the resultant split of the Roles Matrix fields. Fields of the Table are not filled in order to enlighten the principle of the split. Only two fields are filled: the field with roles "knowledge-technology" presents axes, and the field with roles "attitudes-technology" presents the resultant sub-split.

The sub-split of fields in Table 2 drives analysis of expected impacts perturbations in respective TA roles. Given the TA project, the roles are reviewed in order to get better TA. By adding a new sub-split, one allows the investigation of culturally biased evaluations. Thus, the table may serve as an important framing tool. It assists in the identification of the cultural factors for each actor and any cultural differences between them.

Identified differences can – but do not necessarily do so – influence views and require reaction. Hence the evaluation of cultural differences influences concepts, values and views resulting in perception of data, modes of communicating ideas and behaviors.

Table 2 Split of fields in Roles Matrix serves to include cultural factors in TA.

	I. Knowledge	II. Attitudes / opinion		III. Action / initiatives	
technological /scientific aspects	more tangible	spontaneous tangible cultural influence	Controlled tangible cultural influence		
	less->more controlled				
	less	spontaneous intangible cultural influence	- controlled intangible cultural influence		
	tangible				
societal aspects					
policy aspects					

4.2
The method

The approach presented in the Section 1 can be supplemented with a support scenario in which culture influences the identification process (see the Figure 2).

The scenario assumes a sequence of three steps. Given a project, the analyst first identifies the relevance of roles in the specific project setting. Next, the relevant roles identified are listed and for each of them four group of factors (composed of intangible, tangible, controlling, and spontaneous) are formed and reviewed. Finally, specific actions corresponding to each group are identified and evaluated as impact (de)stimulants.

In the second step the analysis (supported by a table of supporting tools) can be built using findings from social psychology (especially taking into account the social influences theory) and behaviuoral decision making theory (especially group judgments). In Table 3 one row of the table clarifies this concept. The row describes the selected role "d" and for this role these general phenomena of social psychology which seem relevant were addressed.

Let us recall shortly the review of definitions (see also Gist et al. 1987). *Social facilitation*. Literature recalls the problem of change in an individual's normal solitary performance occurring in the presence of other people. Performance of simple,

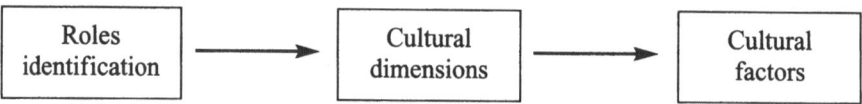

Fig. 2 Identification of cultural factors in TA

Table 3 Examples of findings from social psychology, which drive the search for cultural (de)stimulants in specific cases.

	Intangible	**Tangible**	**Controllable**	**Spontaneous**
Social mapping Role "d" – Policy objectives explored"	Social conformity	Social facilitation	Social comparison theory	Diffusion of responsibility

well-learned responses is usually enhanced by the presence of onlookers[24]. *Diffusion of responsibility*. People do not work as hard in groups as they work alone. Hence the responsibility for the final outcome is diffused among members of a group, whereas individuals working alone have to bear sole responsibility for the outcome. *Social comparison theory*. People often follow others and they are very concerned about the opinion others have of them, which is widely confirmed by research. People evaluate their ability levels and opinions, and compare themselves with others[25]. *Social conformity*. Social comparisons can strongly influence perceptions. It has been proven that factors influencing conformity exist (anonymity). Theory shows that in situations where there are strong pressures to conform, lone dissenters can have a major effect.

Psychological concepts, presented in Table 3, do not directly address role definition and analyzed case. To give an example of how to supplement the general concept with detailed actions, let us return to the example from previous section: the role "d" – "Policy objectives explored". This role is related to the problem of the overall performance of robots. Let us recall the cultural aspects of this example. The robot concept is related with one of the oldest myths about creation of an artificial perfect man. Different presentations of this myth convey emotions: fear and hope. Fear, that robots can dominate mankind. And hope, that robots will – at least – facilitate our life. People may not be conscious of their primary emotions. Faced with professional terminology they may have serious difficulties with formulation of views in this respect, creation of options, criteria definition and sound justification of their positions in the debate. Typical behaviours manifest resistance, avoidance or on the contrary – hyper-optimism.

The role "d" requires exploration of policy objectives. The result of this exploration will lead to better understanding, commitment and wider sharing policy

---

[24] However, the performance of complex, unmastered skills tends to be impaired by the presence of others. This effect – social facilitation – is at least partly due to presence of others. Later research showed that this enhancement takes place even when the prospect of being evaluated by others does not involve their physical presence. Social facilitation was investigated with a variety of verbal and mathematical tasks and real environments.

[25] Central to propositions of the theory stated that: a/ people have a natural tendency to evaluate their opinions and abilities, b/ to the extent that objective, non-social information is unavailable, people evaluate their opinions and abilities by comparison with the opinions and abilities of others, c/ given a choice, people prefer to compare themselves with others who are close to them in opinions and abilities.

objectives, and sometimes – their redefinition. This will be reflected in steps of the TA-structure: including the cost-benefit-analysis, legal aspects and ethical aspects. Social conformity will intangibly influence parties, who will resist values perceived as common, agreed, and implicitly, unquestionable. Different cultures will manifest these attitudes to a different degree. When facing such attitudes in debate about a project, one should identify their emotional roots and manifest respect rather than attempting to directly influence such behaviours. Social facilitation will tangibly influence the debate. Therefore, it seems recommendable to increase public visibility of the debate and the actors. This will increase performance of working groups. Also, in this situation, cultural differences will influence the degree of eagerness to compromise on policy objectives. Social comparison theory suggests use of similar contexts in the argument. Comparison with twin communities and their systems of objectives is a controllable operation and it is compatible with human nature and facilitates identification of one's own objectives. Finally, the process involves diffusion of responsibility, which in a non-controllable, spontaneous way will influence the objectives under review. Here recommendable actions should tend to decrease anonymity amongst the debating committee members. All the roles in the Role Matrix can be reviewed with respect to *tangible-controllable* structure in the same way as presented above for the exemplary role "d", and appropriate operations determined to decrease the undesired outcomes of cultural factors influencing the assessment.

When choosing a strategy, one should be aware of one's own and others' culture, and understand the specific factors in the current relationship, and predict or try to influence the other party approach.

Table 3 can be expanded for all roles in the Role Matrix. It provides impact analysis with a toolbox for analysing the cultural aspects of the problem and building recommendations concerning TA and its impact. This toolbox may serve to create checklists that help in project design and mangement. There is a typical tradeoff between the degree of specificity in building checklists and their direct usefulness. The more detailed the checklist is, the more easily it can be applied, but this also narrows the class of suitable situations.

5
Concluding remarks

The chapter presents a concise briefing of culturally-based factors that influence problem formulation, individual and group decision making. It appears that there is an extensive record on such factors influencing human behaviors – in particular in organisations, which can be risen, stimulated or controlled, sometimes implicitly.

The influence of cultural factors on technology assessment does not seem to be related with the technology itself. It is the technology assessment which is the source of cultural framing bias. Technology assessment is a process of group interactions. This process starts from identification and description of the technology related impacts, then goes on to framing alternatives and criteria followed by continuous redescriptions.

As cases prove, the results of technology assessment may have a wide range of implications and can touch large groups in a society (if not society as a whole). In this situation, culturally rooted factors may lead to a biased assessment. Biased assessment may effect in undertaking decisions which do not correctly compromise stakeholders motives. Still, judged after procedural correctness such decisions may seem appropriate. Thus bias threatens effective decision making and impedes reliable assessment.

Described factors act in any social environment, thus the monitoring of cultural factors is recommended as a recipe for technology assessment itself. However, findings presented in the book, and especially identifiction and classification of TA roles, allow a general approach focused on the technology assessment process to be presented. This approach requires the analyst to follow the procedure of role description in the actual case context. Given description of behaviours relevant to a given role, the analyst should identify tangible and intangible factors influencing these behaviors. Then the analyst can check which factors can be controlled and to what degree. This knowledge can be included in operations related with technology assessment and consequent recommendations.

References

Adler N (1991) International Dimensions of Organizational Behavior. PWS-Kent Publishing

Bazerman M (1991) Judgment in Managerial Decision Making. John Wiley & Sons, New York

Berleman WC, Steinburn TWW (1967) The Execution and Evaluation of a Delinquency Prevention Program. Social Problems, vol. 1, no 4

Deal TE, Kennedy AA (1983) Culture: A New Look Through Old Lenses. Journal of Applied Behavioral Science

Decker M, Grunwald A (2001) "Rational Technology Assessment". In: M. Decker (ed) Interdisciplinarity in Technology Assessment. Springer, Berlin

Earley PC, Northcraft GB, Lee C, Lituchy TR (1990) Impact of Process and Outcome Feedback on the Relation of Goal Setting to Task Performance. Academy of Management Journal, March

Edwards W, Newman JR (1982) Multiattribute Evaluation. Sage Publications, Beverly Hills

Fischoff B, Slovic P, Lichtenstein S (1991) "Knowing What You Want Measuring Labile Values". In: Bell DE, Raiffa H, Tversky A (eds) "Decision Making. Descriptive, Normative and Prescriptive Interactions". Cambridge University Press

Gist M, Locke EA, Taylor MS (1987) Organizational Behavior: Group Structure, Process and Effectiveness. Journal of Management, Summer

Gramlich EM (1981) Benefit-Cost Analysis of Governmental Programs, Englewood Cliffs, N.J., Prentice Hall

Grey RJ, Thone TJF (1990) Differences Between North American and European Corporate Cultures. Canadian Business Review, Autumn

Hofstede GB, Neuijen B, Ohayv DD, Sanders G (1989) Measuring Organizational Cultures Qualitative and Quantitative Study Across Twenty Cases. Administrative Science Quarterly, June

Janis IL, Mann L (1977) Decision Making: A Psychological Analysis of Conflict, Choice and Commitment. Free Press, N.Y.

Keeney RL, Raiffa H (1976) Decisions with Multiple Objectives: Preferences and Value Tradeoffs. John Wiley & Sons, New York

Kluckhohn C (1954) Culture and Behavior. Free Press, N.Y.

Kluckhohn C, Strodtbeck F (1961) Variations in Value Orientations. Greenwood Press, Westport

Knotts R (1989) "Cross-Cultural Management: Transformations and Adaptations". Business Horizons, January–February

Liakopoulos M (2001) "The Politics of Technology Assessment". In: M. Decker (ed) Interdisciplinarity in Technology Assessment. Springer, Berlin

March JG (1988) Decisions and Organizations. Blackwell

Meek VL (1988) Organizational Culture: Origins and Weaknesses. Organization Studies, vol. 9, no 4

Phatak A (1995) International Dimensions of Management, Thompson Custom Publishing

Schein EH (1990) Organizational Culture. American Psychologist

Sherrill S (1984) Toward a Coherent View of Evaluation. Evaluation Review, vol. 8, no 4
Simon H (1976) Administrative Behavior. Free Press
Tversky A, Kahneman D (1991) "Rational Choice and the Framing of Decisions". In: Bell DE, Raiffa H, Tversky A (eds) Decision Making. Descriptive, Normative and Prescriptive Interactions. Cambridge University Press

Part III – Appendix

Appendix A
Description of Examples for Matrix of TA Roles

Example for role "a": Technical options assessed and made visible

Title of the project:
"Open Source Software in digital administration"

Reference/web-link to description of institute:
Teknologirådet – The Danish Board of Technology (www.tekno.dk)

Reference/web-link to documentation:
Teknologirådets rapporter 2002/13, Copenhagen 2002. ISBN 87-90221-76-1 / ISSN 1395-7392: "Open Source Software – i den digitale forvaltning. Analyse og anbefalinger udarbejdet af en arbejdsgruppe under Teknologirådet".
http://www.tekno.dk/pdf/projekter/p02_open-source-rapport.pdf (in Danish)
http://www.tekno.dk/subpage.php3?article=494&language=uk&category=11& toppic=kategori11 (in English)

Short description of project:
Open Source Software (OSS) has been actively promoted in European public administration. The question revolves around how Open Source can best be implemented, and at what tempo. What are the obstacles now, and what other obstacles might turn up? Where and when should an official concerted action involving Open Source be initiated and what advantages and drawbacks will supervene when Open Source is used in the public administration? What have been the experiences with integrating OSS in the existing IT-milieu, dominated up until now by proprietary software? How extensive at the present time is the use of OSS in public administration in Denmark? What obstacles stand in the way of making use of OSS? What would the economic consequences be of a total implementation of OSS in the public sector of Denmark?
An expert working group composed of experts into ICT management, IT economy, and e-governance evaluated the status and experience of OSS implementation in the public sector, and developed an economic model on the socio-economic consequences of a total implementation of OSS in the public sector. The group developed a layout for a strategy for establishment of open standards for document filing and exchange.
  The findings were difficult to ignore, and gave rise to huge attention. If the public sector followed an investment path towards full implementation of OSS solutions through a ten years period, the net socio-economic result would be a yearly public sector saving of the magnitude of 4–6 billion DKK equalling 1000 DKK per capita

(~650 mill. € / 150 € per capita). The investment needed would mainly be connected to staff education and document open standard implementation. The project resulted in a revision of the ICT strategy of the Danish government. The new strategy points out that open standards should prevail in the future, and that OSS should be evaluated seriously in connection to ICT investments in the public sector.

Involved methods:

The Expert Working Group method at the Danish Board of Technology is an open scheme for project design, giving suggestions to the use of certain elements, such as:

- Composition of group, and group roles
- Initial brainstorm and information gathering techniques
- Hearing processes in the synthesis phase
- Evaluation and peer procedures
- Communication activities (conference, seminars, press etc.)

Impacts/roles of projects:

The future-oriented role of "Technical options assessed and made visible" was the main goal of the project when it was designed. The intention was to contribute to the clarification of an issue that was dominated by very rigid standpoints and huge economic interests. The project fully reached that goal, although the strong positions in the debate basically reacted to the analysis, as it would be expected from a stakeholder analysis. The Danish report has been downloaded 50.000 times, ICT professionals and university students being the main users of the analysis.

"Innovations implemented" is an equally important role that this study fulfilled. The acknowledgement of the qualities and economy of OSS solutions has increased considerably among public ICT decision-makers. There has been no evaluation of the direct effect on the use of OSS, and obviously, it is difficult and too early to measure such effects. However, it seems reasonable to state that there will be a long-term effect on the implementation of this technology and widening of choice of solutions, as a result of this project.

Influencing/amplifying factors or initiatives:

Lobbying and a rather aggressive campaign against the project from especially one provider of proprietary software had negative as well as positive influence on the impact of the project. The negative effect had mainly to do with the discussion climate around the release of the report, since it was made difficult for the ICT professionals to discuss the report seriously, without being drawn into a loud fight between the most extreme wings in the debate. The positive effect was connected to the fact that the more aggressive the provider acted, the more people regarded the report as one "that had hit the nail".

An amplifying factor was the fact that there were political forces already waiting for the outcome when the report was published. Some politicians and civil servants already saw the costs and the lock-ins related to the most frequently used software solutions as a major disadvantage with the established policies. These actors could directly use the report, since the analysis supported their main thesis.

Example for role "b": Comprehensive overview on consequences given

Title of the project:
E-Commerce

References/web-link to description of institute:
Institut für Technikfolgenabschätzung und Systemanalyse (ITAS), Karlsruhe
Institute for Technology Assessment and Systems Analysis (ITAS), Karlsruhe
http://www.itas.fzk.de/

References/web-link to documentation:
Riehm, U.; Petermann, Th.; Orwat, C.; Coenen, Chr.; Revermann, Ch.; Scherz, C.; Wingert, B.: E-Commerce in Deutschland – Eine kritische Bestandsaufnahme zum elektronischen Handel. Berlin: edition sigma 2003 (Studien des Büros für Technikfolgen-Abschätzung beim Deutschen Bundestag, Bd. 14)
German summary:
http://www.tab.fzk.de/de/projekt/zusammenfassung/ab78.htm
English summary:
http://www.tab.fzk.de/en/projekt/zusammenfassung/ab78.htm

Short description of project:
At the suggestion of the Bundestag Committee for Economy and Technology and on behalf of the Committee for Education, Research, and Technology Assessment a project on the economic prospects for electronic commerce ("e-commerce") in Germany was carried out by the Office of Technology Assessment at the German Parliament (TAB) between September 2000 and June 2002. The project was performed in close cooperation between staff members of ITAS and TAB. The thematic focus was on the expected diffusion processes, structural changes and their consequences. This includes the elaboration of possible options for political action. In order to reach the necessary comprehensiveness to base recommendations for political options on the following questions have been taken into account:

– What prospects does electronic commerce offer for growth and employment in Germany?
– How important are alternatives to Internet-based electronic commerce, such as electronic commerce via television or mobile phones?
– What production concepts, logistics strategies and infrastructural measures are necessary to react to the requirements of electronic commerce?
– What are the consequences of electronic commerce in terms of ecology (traffic-related aspects, dematerialization, waste prevention, energy and resources efficiency, etc.)?
– What need for legal deregulation or regulation exists to fully develop the potentials of e-commerce in Germany (e-commerce policy)?
Moreover, the above mentioned general questions have been complemented by detailed sector analyses of e-commerce, e. g. for the automobile trade, food trade,

trade with books, sound storage mediums, and videos, trade of pharmaceuticals, electricity, securities, services and public electronic procurement.

One of the major results of the study was that no pressing policy action was needed to ameliorate possible detrimental effects of e-commerce. In contrast, the study emphasised that the e-commerce development was much slower than expected during the "hype" in the years 1998 to 2000, and also the current impacts are moderate (on the job market, for traffic increase etc).

Involved methods:

The project team did an enormous amount of desk research on the broad issues of the project. It was supported by the scientific advice of a couple of external experts who delivered reports on special topics. The results of these reports were discussed in two expert workshops. On the basis of these scientific endeavours the project team compiled the final report including the recommendations for action.

Impacts/role of project:

The report was discussed in the committees and the plenary sessions of Deutsche Bundestag in the usual way. On July 2002 the final report of the project was debated in the Committee for Education, Research, and Technology Assessment and accepted for publication as official Bundestagsdrucksache 14/10006. On December 2002 the plenary session of Deutsche Bundestag assigned Drucksache 14/10006 to eight committees. After debates in these committees, which show quite a good dissemination of the report, the plenary session decided on September 2003 not to deal with this topic again. Therefore the Bundestag followed the main recommendation of the project to follow the issues of an "e-commerce policy" in already established actions like copyright legislation, competition policy, consumer protection, privacy, research policy etc. (see below "influencing factors).

Moreover the report achieved some interest from the scientific community, i.e. in a conference contribution (invited paper), papers, journals and the broadcast. Initial informal feedback indicated that researchers see the study as a comprehensive report of the status quo and future prospect of e-commerce developments in the German economy and in specific industries. (For example: Riehm, Ulrich; Orwat, Carsten; Petermann, Thomas (2002): Stand, Perspektiven und Folgen des E-Commerce, in: Weinhardt, Christof; Holtmann, Carsten (ed.): E-Commerce. Netze, Märkte, Technologien, Heidelberg: Physica-Verlag, pp. 1–18.)

Influencing and amplifying factors or initiatives:

During the project elaboration some main topics connected to e-commerce have been established on the political agenda like copyright legislation or competition policy, which were in the focus of the project as well. For these and other current issues input for parliamentarians was given in workshops and interim reports. In general one could observe that e-commerce was increasingly no longer the main focus of attention of policy, science and the public or had undergone differentiation into specific topics.

Example for role "c": Structure of conflicts made transparent

Title of the project:
Health Risks posed by silicon implants in general with special attention to breast implants, Working document for the STOA Panel, European Parliament, Directorate General for Research, Luxembourg, June 2000, PE 168.396/Fin.St./rev.
Update Report: Health Risks posed by silicon implants in general, with special attention to breast implants, Working document, STOA 116 EN, European Parliament, Directorate General for Research, Luxembourg, June 2003

Reference/web-link to documentation:
http://www.europarl.eu.int/stoa/publi/pdf/99-20-02_en.pdf

Short description of project:
The study was conducted by an external contractor and managed by the European Parliament's Scientific and Technological Options Assessment service (STOA) on a request from the Parliament's Petitions Committee. The latter was responding to two petitions submitted by self-help groups of women patients, who were calling for a ban on silicone breast implants (SBIs), due to the adverse health effects they had suffered following implantation.

The experts (Prof. José Martín-Moreno, Dr. L. Gorgojo, Dr. J. Gonzalez and Ms. W. Wisbaum from Escuela Nacional de la Sanidad, Madrid) were asked to review the scientific aspects of the matter, but to also take into account the views of the different interest groups involved. The purpose of the study was to formulate a series of options for policy-makers. The study was presented by its authors at a meeting of the Petitions Committee, in the presence of representatives of the petitioners and the European Commission. It was published in June 2000.

The study presented three options, including no ban (status quo), a complete ban on SBIs and no complete ban, but adoption of "specific measures to increase and improve information for patients, tracking and surveillance, quality control and assurance and key research". The third option was singled out by the authors, in view of the evidence, as the most balanced.

An update of the study performed by the same team of experts on behalf of STOA was published by the European Parliament's Directorate for Research in June 2003.

Involved methods:
The experts conducted a thorough search of the existing scientific literature (their bibliography contains some 1300 entries, of which the 40 scientifically soundest are cited in the report). They further interviewed 30 patients and obtained information via a questionnaire from 8 self-help groups (plus 8 individual experiences), 9 reconstructive surgeons, 6 silicone breast implant manufacturers, 12 governments, 4 experts in the field and 10 scientific societies.

Impacts/roles of project:
Following completion of the report, the Petitions Committee gathered information from several other committees of the European Parliament (Women's Rights, Envi-

ronment and Research) and the plenary adopted a resolution on 13 June 2001 asking for attention to be focused "on the safety and quality of the products and on pre- and post-operative support" and recommending that "further scientific and clinical research be carried out".

The Commission published a Communication on Community and national measures in relation to breast implants (COM(2001)666) on 15 November 2001. The Communication, which largely reflects the findings of the STOA report, specifies "essential safety requirements and conformity assessment schemes ... in relation to breast implants", as well the "information to be provided to women who are considering receiving breast implants". It also announces a reclassification of breast implants in the framework of Directive 93/42/EEC on medical devices, upgrading them from class IIB to class III products, in the framework of a full quality assurance system. This was finally done via Commission Directive 2003/12/EC of 3 February 2003.

While the issue of SBIs attracted little attention in Europe until the matter was taken up by the European Parliament, several EU Member States have in the meantime taken measures with respect to SBIs. Although these measures differ between countries (mandatory or voluntary instruments, differences as regards registers and advertising), the Commission Communication of 2001 has had an unquestionable catalysing effect.

It is important in this case to recognise that, by carefully delineating the two extreme positions and thus illustrating their ultimate untenability, the STOA report made the compromise "third" option appear all the more natural. Understanding the structure of the conflict in this case helped develop a consensus around the middle ground, in a way that addresses the concerns of the patients (better monitoring, adequate consultation, quality control and more research), without limiting the choice of those who, after due consultation, opt for implantation and while maintaining the manufacturing capacity that will respond to this demand.

Influencing/amplifying factors or initiatives:

The action of self-help groups was clearly essential in awakening public interest in the issue of SBIs. The process was clearly accelerated thanks to the active response of the European Parliament, its willingness to investigate the matter in depth and its ability to formulate specific demands on the basis of the information it obtained. This response was an instrumental amplifying factor in the process that clarified the opposing viewpoints and led to the adoption of concrete legislative measures within a few years after the Parliament took up the matter.

Example for role "d": Policy objectives explored

Title of the project:
Robotics. Options for the replaceability of human beings.

Reference/web-link to description of institute:
Europäische Akademie zur Erforschung von Folgen wissenschaftlich-technischer Entwicklungen GmbH (European Academy for the study of scientific and techno-logical advance)
http://www.europaeische-akademie-aw.de

Reference/web-link to documentation:
Christaller, T., Decker, M., Gilsbach, J.-M., Hirzinger, G., Lauterbach, K., Schweighofer, E., Schweitzer, G., Sturma, D., "Robotik. Perspektiven für men-schliches Handeln in der zukünftigen Gesellschaft". Springer, Heidelberg 2001. ISBN: 3-540-42779-1
http://.www.europaeische-akademie-aw.de/schriftenreihe.html

Short description of project:
Robots have to possess a number of basic skills if they are supposed to be capable of performing actions in the world. One of these skills is the capability of locomo-tion, which is often realized by way of wheels. Alternatively, "natural" ways of movement are copied through the construction of legs, wings, scales, fins and the like. The possibility of perception is implemented with cameras and other sensors, which provide data for modeling the environment. The third important aspect is the capability of learning. The learning of movements, of interpreting the model of the world and of reflection, e.g. with regards to the relation of the robot to its environ-ment, are thereby considered to be the most important areas. Robots which possess these capabilities of acting are called "autonomous robots".

A problem oriented technology assessment has first of all to deal with technical aspects. Are the robots able to do what they are supposed to do? Usually the devel-oper of a robot defines the technical criteria the robot has to reach. In addition to that economical aspects have to be taken into account. Is the overall performance of the robot cheaper than the comparable up to now chosen achievement? This is typi-cally done by an extensively cost-benefit-analysis. Moreover legal aspects have to be considered due to liability aspects. Who is responsible for malfunctions of the robot? Especially, if the robot might by equipped with a learning algorithm? Finally, ethical aspects have to be taken into account because the treatment and the nursing of patients, elderly people, handicapped etc. is deeply rooted in our social behavior and therefore one has to ask in which areas we, as a modern society, should go for the replacement of previous human tasks by robots and in which areas we just do not want a robot act instead of a human being.

Involved methods:
The project has been realized according to the so-called "project group concept" of the European Academy. This is in general an interdisciplinary expert discussion

structured in an introductory phase, during which a common language for the inter-disciplinary endeavour has to be developed, an analytical phase, which can mainly be described as a descriptive stock taking and the development of first argumenta-tion chains, and in a recommendation phase during which these argumentations chains will be expanded into concrete recommendations for action.

This structured discussion process is accompanied by several evaluation loops which ensure high level interdisciplinary research. Different interdisciplinary peer reviews take place at the beginning and the midterm of the project. Moreover the scientific council of the academy, an interdisciplinary expert group as well, scruti-nizes the report at least twice. The outcome is a report in form of a book, including the perspectives of the different scientific disciplines as well as the resulting recom-mendations to act.

Impacts/roles of project:

The result of the project has been summarized in 16 concrete recommendations for action. One of the recommendations to act referred to the problem of liability for damages caused by autonomous robots. It might be that autonomous robots, equipped with learning algorithms, cause a new kind of problem within the liability arrangement of "owner" and "producer". Up to now the machine owner is liable for damages resulting of his or her own defaults. This "malpractice" is mainly deficient organisation, operation or maintenance. The producer of the robot is liable for pro-duction-, construction- or instruction failures. The learning algorithm might cause a problem within this arrangement since the robots learns in its new environment which results in the fact that the robots producer is not able anymore to predict actions of the robot. He or she might therefore refuse liability claims. The robot owner is usually not an expert for robotics or more precisely for learning algo-rithms. This means that the owner does not want to be responsible for actions "learned" by the robot, too.

Since within a working context we have additional accident insurances and work place guidelines the recommendation focuses on unconcerned third parties: People who are affected by the robot while the robot did "its work" for example in a railway station, in a shopping mall or in a pedestrian zone. Here, the recommendation is to go for shifting of the burden of proof. It is not acceptable that an unconcerned per-son has to prove that a complex technical system like a robot caused the damage as it would be according to the current law. In contrast, the robot owner and/or the robot producer have to prove that the system worked well. This avoids that uncon-cerned parties remain with their damages due to irresolvable malfunctions of the robot. The recommendation is accompanied by proposing a liability insurance sys-tem for robot owner which assists them in dealing with damages and avoids finan-cial ruin due to larger accidents.

Influencing/amplifying factors or initiatives:

It was the first time that autonomous robots have been assessed according to their intended and unintended consequences. An amplifying factor for the visibility of the study was the fact that the so call "Bill Joy-debate" warning against the conver-gence of robotics, genetics and nanotechnology took place as the results of the proj-ect were presented to the policy makers, the media and the public.

Example for role "e": Existing policies assessed

Title of project:
Sustainable Development and Innovation in the Energy Sector
Reference/web-link to description of institute:
Europäische Akademie zur Erforschung von Folgen wissenschaftlich-technischer Entwicklungen Bad Neuenahr-Ahrweiler GmbH
www.europaeische-akademie-aw.de

Reference/web-link to documentation:
U. Steger, W. Achterberg, K. Blok, H. Bode, W. Frenz, C. Gather, G. Hanekamp, D. Imboden, M. Jahnke, M. Kost, R. Kurz, H.G. Nutzinger, Th. Ziesemer (2002).*Nachhaltige Entwicklung und Innovation im Energiebereich.*Springer, Berlin
English summary:
http://www.europaeische-akademie-aw.de/grauereihe/GR%20Nr_33.pdf

Short description of project:
Almost every energy scenario is based on trends that would lead to an enormous growth in the demand for energy in the coming decades. Meanwhile, at international conferences among other places, one is concerned with the opposite outlook, a massive reduction of greenhouse gas emissions, especially of CO_2 emissions caused by energy consumption. Experts also point out the political risk of depending on mineral oil and remind us of the fact that resources are not inexhaustible. How can this chasm be bridged? How can we shape a more sustainable energy system from the existing one? Hopes are mostly pinned on technological progress and innovations.

So far, however, there are no specific suggestions concerning the extent to which innovations can really contribute to reconciling ever-growing energy consumption with the limitations referred to, regarding the availability of resources and the environment, and with the structural demands on any energy system.

The study examines potentials of sustainable energy innovations in terms of given environmental policies as well as energy security and reliability. These innovations can solve the conflict between the rising energy demand and the restrictions necessary. The recommendations are based on an analysis of target conflicts and actors in the energy sector. The following strategic measures are central:

- Energy should according to its importance regain a top priority in the political arena.
- Higly targeted subsidies should be given for a limited amount of time to speed up the market introduction of energy-efficient and regenerative techniques in analogy to the "Dutch model".
- Negotiated agreements and unilateral self-commitments can subsequently ensure further market diffusion of sustainable energy innovations.
- The basic research in energy should not be diminished but intensified instead.

Involved methods:

Research Project according to the project group approach of the European Academy

Impacts/roles of project:

A central part of the study report is the evaluation of policy alternatives in terms of their economic effects (role f). On the basis of this assessment a strategy is developed and recommended (role s).

Example for role "f": Setting the agenda in the political debate

Title of the project:

Integral Ethical Assessment Biotechnology

References/web-link to description of institute:

Rathenau Institute

www.rathenau.nl

References/web-link to documentation:

Rathenau Instituut/Stichting Consument en Biotechnologie, 'Integral Assessment Biotechnology. Views from Society' (2003)

Rathenau Instituut, 'Integral Ethical Assessment Biotechnology: Broad Assessment or Trend Analysis?' (2003)

Rathenau Instituut, Letter to Parliament (October 28th, 2003)

Short description of the project:

In January 2002 the Dutch parliament discussed the way biotechnological innovations should be dealt with from a legal, ethical and public point of view. The parliament passed a resolution in which the government was asked to provide an integral ethical framework for assessing biotechnological applications. Some organisations that were involved in the field of biotechnology, however, doubted the desirability and the effectiveness of an integral ethical framework. These doubts induced the Rathenau Institute to start a project to explore these issues. Important questions were: What political needs underlie the parliamentary demand for a framework for integral ethical assessment? What do involved organisations (expert committees, non-governmental organisations, biotechnology companies) think about the current situation, in which the acceptability of biotechnological applications is judged, case by case, by several expert committees? What do politicians and involved organisations expect from an integral ethical framework?

Involved methods:

To investigate how politicians and involved organisations think about the existing practice of biotechnology assessment and what possible improvements can be expected from an integral ethical framework, five members of parliament and twelve representatives of expert committees, non-governmental organisations and biotechnology companies were interviewed in-depth. The results of these interviews were reported in working documents. The persons being interviewed were given the opportunity to comment on the reports. The results of the interviews were analysed in a discussion paper by the Rathenau Institute. The outcomes showed a widely felt need for a broader assessment of biotechnological applications. Two options were put forward to fulfil this need: including ethical and societal aspects within the case by case-assessment, or complementing it with a separate analysis of biotechnological trends that raise moral and societal questions (trend analysis).

On basis of the discussion paper the Rathenau Institute organised a round-table conference in which representatives of involved organisations and some experts, politicians and administrators participated. In this (closed) meeting the two options mentioned above were discussed. The purpose of the meeting was to explore whether the participating organisations and experts could agree upon one of the options. The politicians and administrators were asked to comment on the outcomes. At the end of the meeting it turned out that a majority of participants preferred the second option, the trend analysis.

The Rathenau Institute reported about its findings in a letter to the parliament. Based on the outcomes of the round-table conference, it recommended to provide for a two-yearly analysis of biotechnological trends, in addition to the existing case by case-assessment.

Impacts/roles of the project:
The project was meant to contribute to the political discussion about how biotechnological innovations should be dealt with from a legal, ethical and public point of view. The Rathenau Institute succeeded in informing the parliament about its findings just before the parliament discussed the answer of the government to the resolution. This strategy turned out to be rather successful. The letter in which the Rathenau Institute reported about its findings played an important role in the parliamentary discussion. Moreover, the letter induced the Secretary of State of the Ministry for Housing, Regional Development and the Environment to propose to provide for a two-yearly trend analysis.

It can be concluded that the project of the Rathenau Institute had direct political impact. The recommendation of the institute had been politically addressed and was taken up by the government – to a certain extent. Namely, the proposal of the Secretary of State implied that the existing expert committees should provide the two-yearly trend analysis, whereas the Rathenau Institute in its recommendation argued for a broader design, in which also other involved organisations and the public could participate.

Influencing/amplifying factors or initiatives:
The parliamentary resolution in which the government was asked to provide an integral ethical framework for assessing biotechnological applications had been the starting point for the project. Due to this resolution, the project of the Rathenau Institute gained political relevance. Besides, the resolution offered the institute the opportunity to have impact on the course of political decision making. One of the reasons for having this impact probably has to do with the way the institute repeat-

edly involved politicians in the project: interviewing them, letting them comment on the interview reports, inviting them to participate in the round-table conference, and sending them the letter with recommendations.

Example for role "g": Stimulating public debate

Title of the project:
Dutch debate on clones and cloning

References/web-link to description of institute:
Rathenau Instituut
www.rathenau.nl

References/web-link to documentation:
Biesboer, F. et al. (1999). Clones and cloning: The Dutch debate. The Hague: Rathenau Institute; Working document 70.

Short description of the project:
The announcement of the birth of the cloned sheep Dolly in February 1997 released a storm of emotional reactions and debate all over the world. The Netherlands was no exception. After the initial outcry, the Dutch parliament called for a public debate on cloning and a ban on cloning experiments whilst the debate lasted. In response the Minister of Public Health, Welfare and Sports asked the Rathenau Institute to organise a nation-wide public debate on cloning. At that time the Rathenau Institute had already decided to pay attention to the societal questions surrounding cloning.

The project consisted of two phases. The goal of the first phase was to define the agenda of the debate. Important questions were: What are realistic expectations and what is science fiction? What are the relevant societal and ethical issues? In the second phase of the debate the content of the debate was deepened and participation was broadened.

Involved methods:
To set the agenda of the debate in an open, transparent and legitimate way, a hearing was organised in March 1998. A panel of (former) Members of Parliament questioned researchers, representatives of biotechnology companies and interest groups, and ethicists about the state of the art of the technology, the possibilities of application, the arguments for and against certain applications of cloning and the reasoning behind these arguments. The hearing succeeded in identifying the most important issues for further debate. The results of the hearing were published in popular book, of which a first copy was given to the Minister of Health Care.

Deepening the debate was done by organising expert meetings on cloning of stem cells, cloning of animals for the production of medicine, and animal cloning in husbandry.

Two meetings were held to obtain more insight into the way that various religious and political traditions deal with the ethical problems surrounding cloning. In the debate on cloning and political traditions some computer-animated future scenarios were presented on which politicians and participants could react. To further the public's input to the cloning debate a lay panel was set up. The panel could take part in all of the other activities organised by the Rathenau Institute and was also given the time and money to develop its own activities (e.g. questioning experts, visiting firms, etc.). Also, a survey was held, in which a thousand people were questioned on their knowledge and opinions on cloning. The declaration of the citizens' panel, the results of the survey and the Rathenau Institute's own recommendations were presented to the ministers of Health Care and Agriculture during a final meeting in the Parliament building.

Impacts/roles of the project:

The project was meant to order the hectic and non-informed debate that came about after the announcement of the birth of Dolly, and has been rather successful in doing that. Instead of ready made policy strategies or options the societal debate led to a list of issues and related arguments for further political debate. The comprehensive process made the outcome of the debate an authoritative state of the art description with respect to the involved technological development as well as the related social debate.

The debate had direct political impact for the Ministries of Agriculture and of Health Care. Both ministers were present at the final debate. Since the cloning debate a broader debate around biotechnology has come to existence. With respect to the Dutch Parliament, the attention for this type of topics and the feeling of the need to publicly discuss these technological developments seem to have strongly increased.

Influencing/amplifying factors or initiatives:

An important political event during the first phase of the public debate was the decision by a majority in Parliament that no approval should be given of cloning experiments with animals before the end of the societal debate. The parliamentary decision made the Rathenau debate more politically relevant. It also caused the Ministry of Agriculture to become more interested in the course and results of the debate.

Example for role "h": Introducing visions or scenarios

Title of the project:
Towards a Societal Agenda for Food Genomics

References/web-link to description of institute:
Rathenau Instituut
www.rathenau.nl

References/web-link to documentation:
Van Est, R., L. Hanssen & O. Crapels (2003). *Genes for your food – Food for your genes: Societal issues and dilemmas in food genomics.* The Hague: Rathenau Institute; Working document 92.

Short description of the project:

In the year 2001 the Dutch Ministry of Agriculture organised a public debate on Biotechnology and Food. The debate focused on GM-food, a product that had already reached the market place. Little attention was given within this public debate to the impact of biotechnology on the production and consumption of food in the future. At the same time politics was about to decide on whether to support public funding for an extensive research program on genomics, including genomics research in the field of agriculture, food and nutrition (which we will refer to as food genomics). The Rathenau Institute felt that besides talking about a 'license to produce' GM-foods, the debate should also consider a 'license to develop'.

To address that gap, the project *Towards a societal agenda for food genomics* was set up in 2002 to map and discuss possible applications and social aspects of food genomics research. At that point in time, however, a debate on food genomics was merely non-existent. Interest groups were often not aware of the development and also only a small number of social scientists were involved in that field. Consequently, the debate so to speak had to build up from scratch. Five social scientists were asked to write essays about the societal impact of food genomics. These essays provided the central tool for generating social images or scenarios about the future of biotechnology and food, and for starting a debate about the type of future society prefers.

Involved methods:

The project comprised of several types of activities. A scientific study was done to map food genomics research in the Netherlands, and its potential applications. The study also investigated which societal issues scientists involved in food genomics research thought were relevant to address. Five social essays were written to explore the societal impact of food genomics on the entire food chain. The authors were asked to look into the future but relate this future to historical and current developments. The drafts of these essays were discussed in an expert workshop that confronted the essayists with scientists involved in food genomics research. The final versions were discussed during a working conference with a broad audience, consisting of policy makers, business people, scientists and members of societal organisations. The results of the studies, the essays and the discussions were presented during a public hearing, in which parliamentarians questioned experts and representatives from industry and societal organisations.

Impacts/roles of the project:

The project created various visions for the future impact of genomics on food production and consumption (for example, personalised diets). These types of images were not ready available and had to be developed within the project. The working conference presented the first time in the Netherlands, maybe even in the world, that a wide group of experts and stakeholders discussed about the societal implications of food genomics. The role of the project was to raise awareness about this new technological development and to give an overview of the social and moral issues that play or will play a role in food genomics. The societal agenda and recommendations that resulted from the project do not present the final word, but can be used to give the societal discussion about food genomics further direction, content and shape.

Influencing/amplifying factors or initiatives:

During the project the *Netherlands Genomics Initiative* was established to co-ordinate the research activities on genomics. The government and the Parliament demanded that social issues and concerns would get proper attention in this early stage of genomics research. The Netherlands Genomics Initiative reserved the substantial amount of twenty million Euro (about 5 percent of the total research budget) for research into social aspects and communication. This money will be partially spent on the establishment of a *Centre for Society and Genomics*, which has the task of establishing a societal agenda for genomics. The Rathenau Institute has done the groundwork for such an agenda with respect to food genomics. The project also played an important networking role, by bringing together in an early phase the players that will play a key role in the Centre for Society and Genomics and the public debate in the Netherlands on food genomics.

Example for role "i": Self-reflecting among actors

Title of the project:

Shopping spree in the digital glass-house. Electronic recording and evaluation of customer data. Short version of the TA-study "The transparent customer"
TA Study Report, TA 38A/2000

Reference/web-link to description of institute:

Centre for Technology Assessment TA-SWISS
(Zentrum für Technologiefolgen Abschätzung) www.ta-swiss.ch

Reference/web-link to documentation:

http://www.ta-swiss.ch/www-
remain/reports_archive/publications/2000/38A_kf_glaeserne_kunden.pdf

Short description of project:

The "transparent customer" study is the first in a series of *TA-SWISS* studies addressing the economic, legal, political, ethical and psychosocial effects of the recording of personal data. In more precise terms, this study deals with how customer data is stored by private bodies (via customer cards and in particular electronic trade). The first aim is to gain an initial picture of current and future technology used to obtain and analyse customer data and to show what problems may thus be caused within society, between the customer's private sphere and his or her needs, between differing norms and values, etc. Secondly, the study should reveal possible developments in other fields where personal data is recorded and analysed, as for example in the medical sector or by the authorities.

Involved methods:

The TA-Study on the Transparent Client was done as a classical TA Project. The first part of the study involved examining information available in the literature and on the Internet, while the second part will comprise interviews with experts and relevant companies.

Impacts/roles of projects:

This study has had a very large echo among different groups. The media were very interested in the results of the study, as the privacy issue was a very hot topic. It is quite difficult to know in advance when a topic will raise public attention and it was quite a surprise when different journalists asked TA-SWISS about the study. At this time, the study was finished, but not published (once a study is finished, it can take several months before it is made public). As the media never address an issue on a long period, TA-SWISS decided to release the study quicker than it was originally planned (in November 2000 rather than in January-Februar 2001). At this time, the media interest for the issue was still intact and the study was related in most national and local newspapers.

Apart from the media, other groups showed their interest in diffusing the results of the study:

A "café scientifique" has been organised by an association dedicated to scientific dialogue (Bancs publics) on the theme of surveillance and the study was used as a basis for discussions. A panel of experts were invited to discuss the issues of surveillance with the general public

A consulting enterprise on informatics asked TA-SWISS to present the results of the study during one of its events (a conference).

A consumer organisation (Konsumentenforum) chose the theme of the transparent client for its annual meeting. The TA study served as a basis for the discussions.

Arthur Andersen invited TA-SWISS to present the study at a seminar dedicated to good practices in e-commerce.

Influencing/amplifying factors or initiatives:

This very positive echo was certainly due to the actuality of the theme. The study was published in the right moment, when a public discussion began on privacy. This case study shows how external factors on which a TA institution do not have any control can contribute to the success of a TA project.

However, TA-SWISS reacted quickly and with flexibility to the different demands. For example, it decided to make the report public at a time when the study and the short version were only available in German (normally, a report is made public when the short version is available in all languages). It was also a moment when the TA staff was very busy working on other projects (among them a PubliForum), but indeed succeeded in organising a public event on the study.

Example for role "j": Blockade running

Title of the project:

PubliForum "Electricity and Society" 15 – 18 May 1998 in Berne
Citizen Panel Report, TA 29/1998

Reference/web-link to description of institute:

Centre for Technology Assessment TA-SWISS
(Zentrum für Technologiefolgen Abschätzung) www.ta-swiss.ch

Reference/web-link to documentation:

http://www.ta-swiss.ch/www-remain/reports_archive/publications/1998/ta_29_98_e.htm

Short description of project:

The future of our energy policy was happening at the time. Political aspects were becoming more and more complex: several new bills were discussed in Switzerland or were in preparation (Energy Act, CO_2-Act, Nuclear Energy Act , Ecological Tax Reform). Apart from this legislation, the general public was asked to vote on various popular initiatives (Energy-Environment-Initiative, Solar-Initiative, Moratorium Plus, Opting out of Nuclear Energy). By bringing about thirty Citizens from the whole of Switzerland to participate in a "round table" discussion, PubliForum "Electricity and Society" sought to introduce a new dimension of dialogue into current and future debates, which had been very often characterized solely by facts and figures and by the particular interests of the parties affected. In the course of several days of intensive work, the citizens participating in the PubliForum "Electricity and Society" had the opportunity to contribute their own personal experience and their day-to-day worries and expectations in the hope that these would be taken into consideration when formulating future Swiss energy policy. In addition, the discussions which took place between experts and citizens (and which form the basis of this report) should provide a contribution to a new open public debate on an electricity supply system which will be able to face up to the future.

Involved methods:

The PubliForum method comes from Denmark where the so-called "Consensus Conference" was developed. This is a procedure that enables citizens to partake in the discussions on the effects of new technologies. PubliForum sets up a platform for a new dialogue between the worlds of science, industry and politics as well as those directly concerned and the general public. In this example, the citizen panel has drawn up – based on its dialog with the information persons – a report that represents its viewpoint on future energy policy in Switzerland.

Impacts/roles of projects:

Basically, the main role of this project was to meet our future electricity needs in a sustainable way. In this context, "Sustainable" means that future generations will still have a basis for their existence unaffected by problematical wastes and emissions affecting the climate.

The PubliForum on Electricity and Society, as the first participatory procedure ever organised by TA-SWISS, had acted mainly as an icebreaker, both for TA-SWISS, as the PubliForum's success proved the feasibility of participatory procedures in a multilingual context, and for subsequent PubliFora. Citizens experienced the discussions within the panel as stimulating and constructive, and nobody complained about the multilingual character of the procedure. Citizens also succeeded in writing a comprehensive and differentiated "citizen report" on a highly complex issue.

The PubliForum was also an icebreaker for the idea of participation in TA: politicians, scientists and other stakeholders recognised the possibilities and advantage of a new form of dialogue for scientific and technological issues. Even though some remained sceptical, a majority of politicians and experts interviewed by the external evaluators considered such participatory procedures as a promising tool to integrate citizens in the societal discussion on new technologies. The further PubliFora organised by TA-SWISS confirmed this interest for new forms of dialogue, as politicians or actor groups, formerly sceptical about the idea of letting lay people discuss highly complex policy issues, could be convinced to take part in the project.

Influencing/amplifying factors or initiatives:

The "pioneer" role held by TA-SWISS also contributed to its visibility and, more generally, to the visibility of TA; the press, of course, reported on this arrangement, but many experts and stakeholders knew about TA-SWISS and its activities as they participated in the PubliForum, be it as an expert, a member of the accompanying group or, simply, as an interested member of the public.

Example for role "k": Bridge building

Title of the project:

"PubliForum "Genetic Technology and Nutrition" 4 – 7 June 1999 in Berne Citizen Panel Report, TA P 1/1999 e

Reference/web-link to description of institute:

Centre for Technology Assessment TA-SWISS
(Zentrum für Technologiefolgen Abschätzung) www.ta-swiss.ch

Reference/web-link to documentation:

http://www.ta-swiss.ch/www-remain/reports_archive/publications/1999/ta_p_1_99_e.pdf

Short description of project:

The *PubliForum on Genetic Technology and Nutrition* was the second consensus conference organised by the Swiss Centre for Technology Assessment (TA-SWISS). Its aim was to create a real encounter between the actors involved in the development of genetic technology (i.e. scientists, but also industry, authorities and NGOs) and the public. Interestingly enough, one year before (1998) Swiss citizens had had to vote on a popular initiative demanding the ban of genetic engineering (the so called initiative for genetic protection). The political campaign around this initiative was a first encounter for the different concerned actors, but the fact that the rules of the game did not allow for a win-win situation (citizens had to answer with a yes or no vote) meant that it was difficult to get a real dialogue going. With the PubliForum on genetic technology in nutrition, the rules of the game were changed so as to allow for win-win situations. The integration of ordinary citizens would then provide more knowledge on their wishes, alternative solutions and needs. It would also provide an opportunity to learn about their argumentation patterns: how did they perceive and understand the implications of genetic technology in nutrition, what were their hopes and fears , on which basic values and standards did they judge the issue?

Involved methods:
The PubliForum is based on the method of the "consensus conference" model with the objective of initiating dialogue between researchers, political and business policy makers and the general population to promote transparency and public debate.

Impacts/roles of project:
What effects had had this PubliForum on the public and on politics? The PubliForum was extensively reported on in the media. But its main impact maybe occurred in Parliament where a law on genetic engineering was in discussion (GenLex). The Citizen Panel's proposal for a moratorium was cited by some members of Parliament during the plenary sessions. We do not have records about discussion in Commissions, but informal contacts confirmed that the PubliForum had been used in the argumentation of some politicians. Moreover, TA-SWISS and two participants of the PubliForum were invited in the Parliamentary Commission of the Upper Chamber (State Council) to present the project and its results.

In these debates, the Swiss peasants organisation demanded a moratorium on the production and marketing of genetically modified organisms, in which scientific trial fields could nevertheless be permitted. When making this proposal public, the peasants did not mention that the PubliForum came to the same recommendation. However, the peasants' representative in Parliament explained in a private discussion that he came to the idea of such a "differentiated moratorium" in reading the PubliForum report. At the end, however, the proposal for a moratorium was not accepted by Parliament, but it still significantly contributed to the discussion.

Influencing/amplifying factors or initiatives:
The positive echo of the PubliForum certainly benefited from the actuality of the theme. But the efforts from the part of TA-SWISS also contributed to it. First of all, at the time of starting the project, actors involved in the debate about genetic engineering had been consulted to know whether it was the right moment to launch a project on GMOs and whether the participatory form was adequate. All the consulted actors encouraged TA-SWISS to launch such a project. Then, the most relevant actors in the debate around GMO were invited to be member of the accompanying group supervising the whole project. Representatives of industry, environmental organisations and administrations were all sitting at the same table and somewhat committed to the project. Decision-makers were also invited at the main public events of the PubliForum to give a speech. They had thus the opportunity to learn about the project and, in some cases, they committed themselves to take into consideration the PubliForum's results. And of course, the citizen panel's report was widely diffused to the press, to the politicians and to the TA network.

Example for role "I": Comprehensiveness in policies increased

Title of the project:
"Global Sustainable Development – Perspectives for Germany"

Reference/web-link to description of institute:
http://www.itas.fzk.de

Reference/web-link to documentation:
http://www.itas.fzk.de/deu/fb/nachhalt_akt.htm

Short description of project:
The project was worked on in cooperation with several other institutes of the Helmholtz Association of German Research Centres (HGF). By applying the "integrative concept of sustainable development", elaborated in the project as a reference framework for further analyses and evaluation processes, the present situation in Germany was analysed and evaluated. For this a set of core indicators was chosen to make concrete the basic sustainability rules of the concept. By comparing actual data for those indicators with the targets already set in the political process or proposed in the project the most pressing sustainability deficits were identified, on the total economy level and for several activity fields as well. Furthermore possible development paths until 2020 were viewed regarding some of those indicators in the framework of three future scenarios reflecting different basic societal and political trends. Finally alternative political strategies and measures aimed at the solution of the problems and differentiated according to the scenario trends were described and evaluated. The main conclusions drawn from this were first that an adequate realisation of the integrative concept requires political strategies aimed at solution approaches for several problems simultaneously and second that the attainment of the targets set and the availability of sufficient financial resources requires a comprehensive and fundamental structural redesign of existing tax and charge systems.

Involved methods:
The "activity fields approach" worked out during the project to structure the society and the economy adequately were applied for quantitative and qualitative analyses besides analyses on a total economy level. Explorative scenarios were applied to cope with the several uncertainties regarding trends in future societal development. Core documents and discourses concerning the most pressing problem fields were evaluated to formulate quantitative targets for the several core indicators. For the quantitative analyses in the scenarios the environmental economic simulation model PANTA RHEI was applied e. g. using actual input-output-data for Germany on a national level and adjusted to the special activity fields structure.

Furthermore expert panels were applied as well as participative workshops to identify core problems, issues and strategies in specific technology fields analysed in the project.

Impacts/roles of project:

With the integrative concept of sustainable development the project succeeded in contributing to a more holistic and comprehensive understanding in the German sustainability debate which previously had been focused rather on the separate "pillars" of societal development and in particular on ecological aspects. At least since the core document in the political process and in the agenda setting for sustainable development, the Brundtland report, was published intra- and intergenerational justice and the global perspective have to be seen as the constitutive elements of the sustainability principle. This requires a definition of the principle that can be put into practise, including the essential ecological, economic, social and institutional aspects of development and reflecting the several interdependencies between them. However in the debate so far some of those issues have been missing or at least have not been discussed according to their importance.

More concretely the essential ideas of the integrative concept could be brought in a certain sense into the development process of the German sustainability strategy worked out by the federal government. Compared to the beginning of the process this resulted in a modified, broader conceptual approach of the strategy being not any more oriented according to the development "pillars" but on four cross-dimensional sustainability principles. Besides this other research institutions working on sustainable development issues (e. g. the TA-academy in Stuttgart) have modified their approach during the discourse of the last two years at least in the direction of the integrative concept.

Furthermore the integrative concept or at least parts of it are being applied actually in some current scientific projects e. g. concerned with the development and proving of an integrated local sustainability information system in Leipzig and Halle, with the criteria development for the sustainability assessment of renewable resources technologies in Austria, with the criteria development for the sustainability assessment of stock exchange quoted firms in Austria as well or with a modified definition of wealth on a regional level in Germany.

Influencing/amplifying factors or initiatives:

On the one hand the continuation and increasing of several "non-ecological" problems (e. g. (long time) unemployment, poverty or educational deficits) in particular caused a growing societal sensitivity for the issues, for the amount of financial resources necessary to solve the problems and for the requirement of integrative strategic solutions. All this created an increasing pressure for a suitable scientific and political reaction.

On the other hand the essential elements of the integrative concept have been substantiated and justified in the project in great detail and conclusively with respect to the several discourses and core documents of the last 15 years. The core results of the project were published in two books and were presented and discussed with scientists, politicians and stakeholders at a special conference and will be further discussed in detail at a workshop organised in the near future. With several publications on the conceptual framework and specific activity fields and with presentations on national and international conferences it is furthermore aimed at an increasing publicity in several communities.

Example for role "m": Policies evaluated through debate

Title of the project:
"Ageing Society"

Reference/web-link to description of institute:
Teknologirådet – The Danish Board of Technology www.tekno.dk

Reference/web-link to documentation:
www.tekno.dk (in Danish: working papers by A. Schaumann, L.H. Petersen on demography and work production; briefing notes No. 163, 171, 180)
www.folketinget.dk (in Danish: report ISBN 87-7982-018-2)

Short description of project:
"Ageing society" is a worldwide concept that refers to the demographic development towards a higher average age. It is generally acknowledged that in the short term – during the next 50 years – Ageing Society will be a challenge to the welfare because of the decreasing ratio between citizens inside and outside the working force. However, there is a tendency to over-simplify this development, like when the development is labelled "the elderly burden". For example, the capacity of the future elderly is not necessarily the same as the capacity of the present elderly, and a well-prepared policy-making during the next half-century may be able to compensate for the demographic change. The Danish Board of Technology decided in 2001 to try to initiate a project in cooperation with the Parliament, in order to increase the focus, knowledge and readiness for action among politicians on this important issue. The idea was to establish a "Future Panel" of 20 MPs, which would cooperate with the Board on the planning of four thematic parliamentary hearings, and a conclusion phase, which would result in a cross political paper on the most important questions and problems for future policies to handle. The Board contacted the presidency of the Parliament and speakers from all parties in the relevant committees, resulting in unanimous positive reactions. The project organisation Future Panel was established, and the DBT served as secretariat and project manager for the panel.

The DBT arranged for the production of two over-view papers. One paper described the Danish demographic development historically and prospectively, and examined the relation between population, labour force and amount of work. Another paper described the socio-economic consequences and possible options. The papers were used as back-ground information for the hearings. A seminar was arranged, during which the Future Panel accepted the priority list and jointly developed an inventory of 24 conclusions and policy options. By the end of 2003, the Parliament published a book, comprising a) the 24 conclusions, b) the DBT resume and priority list, and c) resumes of the four hearings. The book was distributed to all MP's, the counties and the municipalities, followed by a letter from the President of the Parliament, in which Danish politicians are requested to take up the challenges of the Ageing Society in due time. There has been a request for information and further activities after the end of the project. The municipality of Odense has asked the

DBT to co-arrange a series of three local citizen hearings, focusing on local senior citizen policies. These hearings are expected to be held in autumn 2004. The Committee of Social Affairs of the Parliament has asked the DBT to arrange a seminar between the committee and the advisory group of the project, in order to discuss future social policies in the light of the Ageing Society. The seminar will take place in March 2004.

Involved methods:

The Future Panel method was developed, as a new method at the Danish Board of Technology comprising a Future Panel of MPs, an advisory group, a series of Parliamentary hearings and a Future Panel seminar. The Future Panel method has only been used once. However, the Parliament has expressed satisfaction with the project and openness towards repeated use of the method.

Impacts/roles of project:

The main results of the project were a profound increase of political debate, raising of knowledge, and a new comprehensiveness in policy-making combined with re-assessment of policies. The most prominent role of the Future Panel project is definitely "Policies evaluated through debate". "Policies evaluated through debate" is probably the most important role with an issue as Ageing Society. The long-term strategic development that is needed if such an issue is going to be taken properly care of can only be reached if the full range of actors – and especially political actors – can see the point in those policies that demands action from their side. The Ageing Society project played this role in several ways. First, the project included four Parliamentary hearings, at which knowledge was given, and different normative positions as well as political options were debated in public. The policy fields of relevance were scrutinized at the hearings, and known policies were evaluated in the process. The participants of the hearings included experts; NGO's; institutional representatives; public authorities; local, regional and central politicians.

Second, the project resulted in a consensual policy-inventory made by all parties of the Parliament. All policies were not fully backed up by all members of the panel, but they were brought forward in consensus in order to give a broad scope of options for the actors involved.

Influencing/amplifying factors or initiatives:

The involvement of a recognised policy-maker in the Advisory Group strengthened the respect and the political attention of the project considerably. Former president of the National Bank of Denmark, Mr. Erik Hoffmeyer, was an engaged and invaluable member of the group. The institutional position of the DBT may have been of importance with regards to the setting up of the project. The project could possibly not have been established by the Parliament itself, since it would demand a new formal procedure to be established inside the Parliament. However, the direct link between the DBT and the committees of the parliament made it possible to establish a relation of cooperation, in which the DBT was responsible for the project as such. However, the personal support from the president of the Parliament was crucial for the establishment of this relation.

Example for role "n": Democratic legitimisation perceived

Title of Project:

Online forum on the draft Communications Bill – UK Parliament

Reference/web-link to description of institute:

United Kingdom Parliamentary Office of Science and Technology
http://www.parliament.uk/parliamentary_offices/post.cfm

Reference/web-link to documentation:

http://www.parliament.uk/parliamentary_offices/post/pubs.cfm

Short description of project:

In spring 2002, the UK Government published a draft Communications Bill, setting out its plans for changing the way the media and telecommunications industries are regulated in the UK. As was expected, the draft Bill received a great deal of media and political attention. To handle this in the most effective way, the government authorities in the House of Commons proposed that the bill be examined by a special joint committee with members from both the upper and lower chambers (the House of Commons and the House of Lords), to report by early August. The House of Lords accepted this proposal. The Committee received over 200 pieces of written evidence and also cross-examined a number of witnesses orally. Driven by the subject matter, it was keen to encourage the widest participation in its discussion, so it introduced two innovations: An online forum was commissioned from the Parliamentary Office of Science and Technology and the Hansard Society. This was the first UK Parliamentary online forum to consider a specific piece of proposed legislation, and the first with explicit, on-going interaction with a Committee. All evidence sessions were web cast live and shown later on the BBC Parliament digital TV channel. The online forum ran for a month in June and July 2002, and allowed anyone who was interested to register, to post submissions and to discuss issues with others similarly registered. Committee members from each of the three main political parties in the Houses of Parliament wrote 'think pieces' to encourage discussion in the first few days. The Committee also submitted questions for consideration by those registered at the start and during the forum. Messages were summarised weekly and reported to the Committee at its meetings. Summaries were also placed on the forum website, so that new users could quickly get up to date with the discussion. By the end of the exercise, 373 participants had registered, with 222 messages posted by 136 participants. Some of the issues raised in the forum were included by the Committee in their oral questions to witnesses. The final report referred 22 times to the online forum. In contrast to much written and oral evidence, the forum heard less from large businesses, tending to focus more on the views of individuals, small businesses and pressure groups.

Involved methods:

Mainly desktop research, and organisation and moderation of the on-line forum.

Impacts/roles of project:
It is difficult to disentangle the results of the forum from those of the wider Committee inquiry. Many of the Committee's 148 recommendations (of which 120 were accepted by the government) covered issues raised in written and oral evidence as well as on the online forum. This led to suggestions that future online exercises should focus on specific issues not addressed by the wider inquiry, in particular on people's personal experience. After the conclusion of the consultation, Lord Putnam, the Committee Chair, commented, "The online consultation worked exceptionally well, and proved its worth as a vital tool in the democratic process... The responses were of a very high quality, and gave us a real sense of public opinion across a wide range of issues." Of forum participants who responded to the post-consultation survey, more than 70% thought that the online consultation had been worthwhile, and 91% felt that Parliament should run more online consultations in the future (*Hearing Voices: The experience of online public consultations and discussions in UK governance*, Dr Stephen Coleman, Nicola Hall and Milica Howell, Hansard Society, Oct. 2002).

Example for role "o": New action plan or initiative to further scrutinize the problem at stake

Title of project:
Multi Media – Myths, Opportunities and Challenges (Multi media – Mythen Chancen und Herausforderungen)

Reference/web-link to description of institute:
Office of Technology Assessment at the German Parliament (TAB) (Büro für Technikfolgen-Abschätzung beim Deutschen Bundestag) www.tab.fzk.de

Reference/web-link to documentation:
TAB-working-paper No 33 (German)
Ulrich Riehm und Bernd Wingert (1995): Multimedia – Mythen, Chancen und Herausforderungen" Bollmann Verlag, Mannheim, Oktober 1995, ISBN 3 927901 69 5
English summary: http://www.tab.fzk.de/en/projekt/zusammenfassung/AB33.htm

Short description of project:
Multimedia i.e. interactive access to the integrated package of static material (e.g. text) and dynamic material (e.g. film) as a combination of highly disparate expectations in both commercial and social policy ("information society" as the buzzword) was still in the making at the mid of 1995 and the internet as a new integrated world wide communication technology was at the outset. After preparatory work in the framework of TAB on the issue of multimedia, the Committee for Research, Technology and Technology Assessment commissioned TAB in July 1994 to carry out a preliminary study on a "Multimedia" TA-project. Since multimedia at that time was neither a clearly-defined technology nor a well-defined field of application the project was aimed at determining application areas and technological configurations and doing the first analysis of the various fields. On this basis, barriers to and

potential for further development were identified, options presented, need for action formulated and proposals for a main TA study submitted.

The study closed with some fundamental considerations relating e.g. to the role of the state in establishing a new multimedia infrastructure. Which communication model should such an infrastructure adopt? The basic alternatives identified were: central or decentralised network concept? Equal status for all participants, or broad retention of the originator-recipient model?

Involved methods:

The project was done mainly by desktop research. A set of studies on particular fields of application were commissioned to outside experts.

Impacts/roles of project:

Immediately after completion the TAB report was intensively discussed in the Committee for Research, Technology and Technology Assessment. The report was appreciated as a starting point for political debate on a new and important field of technology development with high relevance to the German economy. The report was published as a printed matter of the parliament and was discussed in several parliamentary committees and the plenary. As a result of these discussions the parliament referring to the findings of the TAB-report decided – instead of commissioning a main TA-study – to set up a parliamentary inquiry committee on the policy aspects of the issue of multimedia. The enquiry committee "Future of Media in Economy and Society" started working in December 1995 drawing on the TAB-report for definition of its main fields of work. In its final deliberation on the findings of the TAB-project the Research Committee unanimously passed a recommendation to the government to further scrutinize possible impacts of the development of information society and to prepare for a legal framework of new I&C services.

Influencing/amplifying factors or initiatives:

Good timing with regard to the policy making process: The project took advantage of the fact that the issue of the world-wide-web was at the horizon, its relevance for policy making however was not yet discussed. The project could contribute to raising awareness for the political implications of this new technology among parliamentarians.

Example for role "p": New orientation in policies established

Title of project:

Research Policy for Sustainable Development

Reference/web-link to description of institute:

Office of Technology Assessment at the German Parliament (Büro für Technikfolgen-Abschätzung beim Deutschen Bundestag) www.tab.fzk.de

Reference/web-link to documentation:
TAB-working-papers No 50 and No 58 (German)
English summaries:
http://www.tab.fzk.de/en/projekt/zusammenfassung/Textab58.htm
http://www.tab.fzk.de/en/projekt/zusammenfassung/AB50.htm

Short description of project:
On behalf of the Committee for Education, Research and Technology Assessment of the German Parliament the project investigated what demands the guiding principle of sustainable development places on research and technology policy. A set of criteria for sustainable research and technology development were elaborated as an orientation framework for a pertinent reorientation of R&D policy and recommendations were given for the set up of a German programme to promote sustainable research and technology as a cross cutting effort by all ministries for all fields of technology.

Involved methods:
The project was done mainly by desktop research on concepts of sustainable development developed so far. As an exemplary case study, the experiences made by the Dutch R&D programme Sustainable Technology Development were assessed. In addition a workshop with international experts on the issue has been organised at the Parliament.

Impacts/roles of project:
The TAB report – Printed Paper "Drucksache 14/571" – was debated together with the report of the Study Commission "Protection of Humankind and the Environment" (Enquete-Kommission) on 25 March 1999 in the plenary session of the German Parliament, and was appreciated by all parliamentary parties as a major conceptual contribution to the "paradigm change" for research policy demanded by the Study Commission. Even while the project work was underway, the Study Commission incorporated the criteria developed by TAB (problem-oriented interdisciplinarity, stakeholder orientation, etc) into its considerations on sustainability-oriented innovation policy. The TAB report also played a major role in the research ministry's conceptual considerations on mainstreaming the principle in promotion policy. Two interpellations by parliamentary parties in 2000 on the subject of sustainability in research policy (one "minor interpellation" by PDS (Socialists), one major interpellation by the Social Democrats and the Green party) referred to findings from the TAB-report when asking for information on the governments activities in the field. The replies by the Federal Government to the interpellations clearly indicate that the criteria formulated by TAB were taken up by the Research Ministry re-orienting research policy in line with the principle of sustainability (Printed Matters 14/2857 and 14/6959). In the March 1999 plenary session mentioned above, the report was referred to by the Committee for Education, Research and Technology Assessment (lead), the Committee for Economics and Technology, and the Committee for Environment, Nature Protection and Reactor Safety. Even though no recommendation for a decision was being given on the TAB report, it is clear that it contributed to an intensive exchange between parliament and the government on the

goals and concepts of a sustainable R&D policy. At a meeting of the Education and Research Working Group of the SPD parliamentary party with the parliamentary state secretary in the research ministry and representatives of the pertinent departments in that ministry, the TAB report was intensively discussed in regard to possible conclusions for ministry policy. Here (as in the cited reply to the interpellations) the establishment of Germany's own cross-cutting programme, as recommended by TAB, to promote sustainable innovations in line with the Dutch model of the Sustainable Technology Development Programme was the subject of an exchange of opinion on suitable organisational steps towards a sustainable R&D policy. The ministry indicated that the criteria presented by TAB had already been taken into account in the individual promotion programmes, and therefore it was not held to be necessary to initiate a particular cross-cutting programme. At the opening conference for a new research programme "socio-ecological research" the secretary of state referred to the work of TAB as having been helpful for the design of the programme.

Influencing/amplifying factors or initiatives:

Good timing with regard to the policy making process: The project took advantage of the fact that the issue of sustainable development had gained some kind of prominence in policy making when the project was carried out (end of the 1990s).

The success in terms of impacts was mainly due to the fact that the project found supporters among the parliamentary group of the social-democrats and that the new minister for research were striving for profiling R&D policy with regard to sustainable development. SD had been stressed as a guideline for the new government in the treaty between the coalition parties (Social Democrats and Greens).

Example for role "q": New ways of governance introduced

Title of the project:

GMO's on the fields?
Pilot experiences of two civil panels on the communal level – a dialogue between science, technology, politics and society.

Reference/web-link to description of institute:

Fondation pour des Gérerations Futures
http://www.fgf.be

Reference/web-link to documentation:

http://www.fgf.be/index2php?section=page&ID=110

Short description of project:

Early in 2002 the refusal of a number of licences for experimental GMO crop fields by the Belgian Federal Minister for Public Health because of the risk of cross fertilisation of nearby non GMO crops led to a public discussion on the efficiency of the applied authorisation criteria. Subsequently the decision was taken to review the existing criteria and explore the possibility of applying sustainability criteria besides risk criteria. Together with this the minister asked for a public participation study to invite civilians to express their opinion and formulate an advice to the minister on the authorisation criteria of experimental and/or commercial GMO crop fields. This would be done at the local community level. The study was commissioned to the Fondation pour les Génération Futurs that already had an experience in organising public participation activities. Two pilot experiments were designed, one in a Flemish commune Beernem and one in Wallonia, in Gembloux. Both would have the same objectives and the same organisational approach but were held in the respective regional language Dutch and French.

Involved methods:
The method used is called panel de citoyens/burgerpanel (citizens panel). It consist of a one day event that has some of the characteristics of the Danish style 'Consensus Conferences'. The procedure enables citizens to go into dialogue with scientists as well as representatives of industry and politics. Based on this, both the citizen panel drew up a report that represents their respective viewpoint on the authorization criteria for experimental and commercial GMO crop fields.

Impacts/roles of projects:
The main role of both these pilot projects was, for the first time in Belgium, to invite citizens to give advice in a structured way on matters of content to implement governmental decisions in a field were traditionally expert opinion dominates the external input in the decision making process. The fact that a local approach was chosen by involving local communities in both the Flemish and Walloon region recognizes the importance that the federal authorities attach to the problems concerning local acceptance of GMO crop fields. This means that in managing governance structures – in which traditionally expert advice plays a mayor role – new forms of participation were explored where specific sensibilities (local resistance to GMO crop fields) were at the core of the process in order to broaden the spectrum of traditional 'science-based' criteria.

Influencing/amplifying factors or initiatives:
The fact that this is the first time the Belgian Federal Government has taken such an initiative means that the lack of experience in implementing the results of such participatory projects could have an influence on its general impact. This could be amplified by the discontinuity of governmental policies in this field. Since the initiative was completed (May 2003), the government has changed (June 2003). In the public and political sphere the initiative was generally perceived as one being closely linked to the views of the responsible (Green) minister, the possibility that the present government (without any Green party presence) changes its policy, is real.

Example for role "r": Initiative to intensify public debate taken

Title of the project:
'New impulses for a societal debate on genetically modified food'

Reference/web-link to description of institute:
Flemish Institute for Science and Technology Assessment (viWTA)
http://www.viwta.be

Reference/web-link to documentation:

http://www.viwta.be/content/en/prj_Current_projects.cfm

Short description of project:
On September 25th 2001 a hearing was held in the Flemish Parliament to discuss the advice published by five Flemish advisory bodies, by request of the Flemish Parliament, on the topic of genetically modified organisms (GM organisms). A recurring element in the five reports was the importance of organizing a public debate on this topic. The Vlaams Instituut voor Wetenschappelijk en Technologisch Aspectenonderzoek (Flemish Institute for Science and Technology Assessment, viWTA, established by Decree on 17/07/2000) provided the opportunity to respond to this advice. The Board of the viWTA decided in December 2001 to organise a pilot project on this topic. In April and May 2002 this topic was narrowed down to 'genetically modified food'.

Involved methods:
ViWTA used a multi-method approach that combines several subprojects that each target different societal groups and stakeholders in order to reach an audience as broad as possible. Different activities took place: a preparatory study on the trends in public perception regarding GMO food in Europe and Flanders; an international symposium where international experiences regarding public participation on GMO food were compared, a public forum on GMO food as the core public participation event and a concluding stakeholder forum where the respective results of the subprojects were presented to the key stakeholders. As a method for a participatory approach the viWTA selected the Danish method of the 'consensus conference'. The viWTA preferred the term 'public forum' and did not retain the obtaining of a consensus in the conclusions and recommendations as an absolute goal for the citizens' panel.

Impacts/roles of projects:
The principle role of this project was to stimulate and structure the debate on GMO food in Flanders, inboth the public and the political sphere. Therefore a multi-method approach that combined several subprojects was chosen, targeting different audiences with different means. In order to reach the main political actor and the client of viWTA, the Flemish Parliament, the results of the public forum were

handed over to the Speaker of the Parliament. He decided to accept the recommendations and suggested to the study should be used as the basis of a resolution to the Flemish Government, asking it to implement these recommendations. Subsequently this initiative was taken over and a proposal of resolution was submitted to the responsible committee, signed by all parties, except the extreme rightist party. Specific recommendations to the Federal Biosafety Council were also taken into consideration and are now being implemented.

The overall role of the project was to stimulate the societal debate among the different actors in the field. This was done throughout the project in the different subprojects where ample opportunities were given to the stakeholders to discuss in a systematic and structured way the problems concerning GMO food.

Influencing/amplifying factors or initiatives:
The experience in Flanders and Belgium regarding the organisation of public participation events in the field of science and technology is rather limited. This is mainly due to the fact that there was no institutional framework to do so, but also due to a limited interest among societal stakeholders and politics. Surprisingly during the year 2003 three main public participation exercises were held: two on the federal level (for one of them see role u) and one on the Flemish regional level, the public forum discussed here. These three events clearly had an influence on each other and influenced positively the external perception on the use and functions of public participation in stimulating public debate. There was some internal consultation among the organising staff in order to guarantee a degree of methodological coherence and non conflicting timings. The public and press interest was also raised by the different events, but the 'surprise' effect of the accumulation of different events all in one year might have also had some dysfunctional effect on the amount of press interest.

Example for role "s": Policy alternatives filtered

Title of project:
Technology Assessment: "Sänger" space transport system (TA – Raumtransportsystem Sänger)

Reference/web-link to description of institute:
Office of Technology Assessment at the German Parliament (TAB) (Büro für Technikfolgen-Abschätzung beim Deutschen Bundestag) www.tab.fzk.de

Reference/web-link to documentation:
TAB-working-paper No 17, 1993 (in German)

Short description of project:
At the beginning of the 1990s a decision had to be taken on how to continue with a governmental research program for hypersonic spaceflight technology (HST), the major part of it consisted of development of a reusable space shuttle system (named "Sänger", after a German engineer and space flight technology pioneer). In 1991 the Committee for Research, Technology and Technology Assessment of the Parlia-

ment commissioned TAB to conduct a study to explore the ways to continue with the HST program. TAB carried out an extensive analysis of the technical-feasibility of, the future demand for and possible impacts of the "Sänger" technology. The study came to the conclusion that a decision on the HST program should involve a general decision on the extent of Germany's future engagement in space flight – since continuation of the HST program was only regarded as being reasonable in the context of an extended German space flight program. Apart from this the study developed three options for continuation of the HST program. One of these options (option III) proposed to expand the scope of the HST program to technical options alternative to the space shuttle technology, to base the program on a systematic comparison of different reusable transport technologies, to intensify international co-operation and to reduce the activities for the development of the "Sänger" technology.

Involved methods:
systematic analysis of technical and economic aspects of the technologies under consideration;. broad scope of arguments pro and con space flight considered, involvement of high level experts from different disciplines.

Impacts/roles of project:
In January 1993 the Committee for Research, Technology and Technology Assessment unanimously forwarded a recommendation for a decision, that the government should restructure the HTS program according to option III of the TAB study, and should enter into consultation with the European partners on the scope and funding of future European engagement in space flight. In March 1993 the TAB study was debated in the plenary and the recommendation given by the Research Committee was approved. When in December 1996 the government was officially asked on the state of reorganising the HST program by the Research Committees group of rapporteurs for TA, the government answered that since 1993 a restructuring of the program according to the decision of the parliament and the findings of the TAB study was under way, and that the activities to develop the "Sänger" technology had been reduced.

Influencing/amplifying factors or initiatives:
Trust in excellence and impartiality of the analysis was decisive for the final adoption of the options developed by TAB. This was prepared for by a well balanced choice of the experts involved and by taking into account a broad range of economic, ecological and even cultural arguments pro and con human spaceflight technology in general and the "Sänger" space shuttle program in particular.

There was a widely recognised political need to clarify the future of the "Sänger" program with regard to its relation to ongoing plans for a European space flight program and with regard to the immense financial challenges involved.

Example for role "u": New legislation is passed

Title of the project:
"Mapping the Human Genome"

Reference/web-link to description of institute:
Teknologirådet – The Danish Board of Technology
www.tekno.dk

Reference/web-link to documentation:
TeknologiNævnets rapporter 1989/6, Copenhagen 1989. ISBN 87-89098-36-6 / ISSN 0903-2789 (in Danish):
"Hvordan skal vi anvende den øgede viden om menneskets gener? – Slutdokument fra konsensuskonference 1.-3.november 1989 arrangeret i samarbejde med Folketingets Forskningsudvalg"
("How should we use the increased knowledge on the human genome? – Final document from the consensus conference November 1-3, 1989, arranged in cooperation with the Research Committee of the Danish Parliament")

Short description of project:
By the end of the 1980'ies, the Danish public debate on gene technology was at its' highest. The American human genome mapping project had begun, giving rise to both wild optimism and deep pessimism about the future prospects for health, economy, societal relations and ethics. The European Union was considering a European genome project. To proactively meet the possible challenges, the Danish Board of Technology (at that time named TeknologiNævnet) established the second Danish consensus conference, which made use of a jury panel of citizens. The project focused on the future situation where an increasing amount of information on a genetic level would be made available to society, and in some cases to the individual citizen. For example, information on risk of diseases, genetic profiles, and gene tests. How should we use this information i.e. in prenatal diagnosis, screening of the adult population, crime investigations, or on the labour market? The consensus conference method was chosen because of the need to gain a comprehensive and qualitative insight into the reasoning of "ordinary citizens", and in order to introduce an unbiased assessor (the citizen panel) into a discourse that was dominated by strong positioning and the influence of professional or economic interests. The consensus conference was arranged by a planning group of expert from different disciplines, together with a planning group of 6 MPs of the Research Committee of Folketinget – the Danish Parliament.

Involved methods:
The Consensus Conference is a well-documented process, which can now be seen as one of the standard methods in the cluster of participatory TA methods. This was the second time it was used in the variant introduced by the Danish Board of Technology, in which the jury consists of non-involved and non-expert citizens. A demographically mixed panel of 15 citizens was established through a newspaper adver-

tisement call, and a panel selection overviewed by the planning groups. The panel met for two preparatory weekends during which they prepared a catalogue of questions that they wished to explore at the conference. An expert panel of 15 experts, covering the necessary scientific areas, and at the same time the different positions in the debate, was established by the planning groups, so that the full question catalogue was covered. The conference lasted 3 days. On the first day, the expert panel gave presentations during which each expert covered specific parts of the question catalogue. On the first half of the second day, the citizen panel and the expert panel had a dialogue about unresolved issues and follow-up questions. The rest of the second day (and night) the citizen panel wrote the final document in which they gave their own answers to the question catalogue. On the third day, the citizen panel published the document at the conference and discussed their findings with the expert panel and the audience.

Impacts/roles of project:

The reason of this example is described in a TAMI context, despite that the project was made nearly 15 years ago, is that it makes up a rare but genuine example of documented long-term impact of a TA activity. The role "new legislation is passed" is the most dominant role of this project, since the consensus conference directly resulted in the establishment of a law on the use of health test information. The parliamentary negotiations on this law directly made reference to the consensus conference as the source of inspiration. In the final document, the citizen panel pointed at the risk of new kinds of discrimination if genetic testing can be used in connection to employment, insurance drawing or reimbursements after work related injuries. However, the law was passed five years later, in 1995, mainly because as the problem grew, the MPs looked into it. more It began as a bill on genetic tests, but ended up as a law on health tests in general. "Comprehensive overview of consequences given" was also a prominent role for this consensus conference, as for most CCs. The question catalogue covers the full range of problem areas, and in the final document they are presented and discussed.

Influencing/amplifying factors or initiatives:

The timing of the project must be seen as an important factor for the relative success of the project. The cooperation with the Research Committee of the Parliament has without doubt increased the impact of the project in many ways. It made the press aware of the event, and the MPs regarded themselves as main users of the results from the beginning of the project.

Appendix B
Description of Institutes

Europäische Akademie zur Erforschung von Folgen wissenschaftlich-technischer Entwicklungen Bad Neuenahr-Ahrweiler GmbH

Mission

The Europäische Akademie is concerned with the scientific study of the consequences of scientific and technological advance for individual and social life and for the natural environment. The main focus is to examine foreseeable mid- and long-term processes that are especially influenced by natural- and engineering sciences and the medical disciplines. As an independent scientific institution, the Europäische Akademie pursues a dialogue with politics and society.

The Europäische Akademie bases its work on the assumption that the sciences have the task, beyond providing specialised scientific information, of also making orientational knowledge available. Therefore, an inter-disciplinary approach is required, bringing together the results from natural sciences, engineering sciences and medical disciplines with thematically relevant studies in philosophy, jurisprudence, economics and social sciences. In addition, the foreseeable results of research and development are to be related trans-disciplinarily to the expected societal needs and positions.

Within its work, the Europäische Akademie takes up and develops approaches of technology assessment, ethics of technology and medical ethics.

Institutional setting

The Europäische Akademie was constituted as a non-profit corporation with limited liability in 1996. Its partners are the Bundesland Rheinland-Pfalz and the German Aerospace Center (Deutsches Zentrum für Luft- und Raumfahrt).

Managing Director is Pofessor Dr. phil. Dr. phil. h.c. Carl Friedrich Gethmann, who is a full professor of applied philosophy at the University Duisburg-Essen. The Federal Ministry of Education and Research (Bundesministerium für Bildung und Forschung) contributes to the financing of the Europäische Akademie in the form of project funding. The Partners' Assembly has formed a Managing Committee to which two representatives of the partners belong. It also nominated a Scientific Advisory Board which supports the Europäische Akademie in elaborating scientific goals and projects as well as in evaluating research results. The members of the different working groups form the Council (Kollegium) of the academy which is a forum for scientific exchange on subjects concerning the activities of the Europäische Akademie.

Methodological approach

The work of the Europäische Akademie is mainly conducted by temporary interdisciplinary project teams. Members of these project groups are recognized scientists from universities and non-university research organisations in Europe. In specific cases, representatives from other areas of society, e.g. from industry, can also be nominated. From their group, the members elect a chairperson.

Trans-disciplinary issues, relevant for several or all project groups, are dealt with in the study groups by staff of the Europäische Akademie.

Through the participation of European scientists, European cooperation with corresponding advisory bodies, through participation in relevant networks and through specific regard to European aspects within the projects, consideration of the European perspective is ensured.

The results of the project- and study groups in the form of subject-related studies are made available as a basis for advice and decision- making. The work is also accompanied by conferences.

Target groups

The results of the Europäische Akademie's scientific work address the relevant scientific disciplines, furthermore political authorities in the fields of science and technology in Europe. Beyond that, with its work it addresses the public, potentially affected by the consequences of science and technology.

The Europäische Akademie expects its scientific work to have repercussions on the conception of science itself. Through rational reflection about the consequences of science and technology, the scientist's responsibility as an important factor in the self-regulation of the scientific system shall be strengthened.

Contact

Europäische Akademie GmbH
Wilhelmstr. 56
53474 Bad Neuenahr-Ahrweiler
Germany
Phone: +49/ 2641 973 300
Fax: +49/ 2641 973 320
http://www.europaeische-akademie-aw.de

THE PARLIAMENTARY OFFICE OF SCIENCE AND TECHNOLOGY (POST)

Mission

POST's role is to assist Parliament, its Select Committees and individual Parliamentarians over issues with a scientific or technological (S&T) content and in their scrutiny of government policies related to S&T policy specifically, and any area with an S&T element more generally. It provides factual briefings and issue analyses that are timely, independent and balanced. POST has a specific remit from Parliament to inform it about developments in S&T related public dialogue procedures. Beyond this, POST has a remit to promote the good image of Parliament in the UK and wider afield by being seen as a competent service.

Institutional setting

POST, highly unusually, is an Office of *both* Houses. The House of Commons covers 70% of POST's running costs, the Lords the remainder. It is administered by the House of Commons. POST was created as a charitable foundation, outside of Parliament, in 1989. Following this successful demonstration, it was taken into Parliament in 1992. In 2001, both Houses established it as a permanent institution. POST has an established staff of nine (director, six science advisers and two administrators/secretaries). It also has well-developed fellowship/internship schemes linking to learned societies, research councils and individual research centres – at any one time about 15 persons are at POST. POST's work is supervised by its Parliamentary Board – members of the Houses of Commons and Lords, and four external representatives from the fields of science, medicine and engineering.

Methodological approach

POST has its own programme, determined by its Board. Outputs are either short (2–8 page) briefings (the majority) or longer reports, produced by 'classical' expert analysis. Some analyses include inputs from various consultative techniques (see below). Sometimes, (e.g., its work on flooding) POST supplements its analysis by factual information for each Parliamentary constituency. For larger projects, POST may convene an advisory committee, usually chaired by a Board member, with business, government, academia and pressure group representatives. POST's committee work has a wider range of outputs, from briefings prior to the start of an inquiry, through analysis of material received while underway, to follow-up work afterwards. A key feature is that all POST's output receives wide-ranging external expert review before publication. POST has pioneered use of consultative techniques in the assessment of S&T issues at Parliament, in particular internet consultations. POST was on the advisory committees of the two UK national 'consensus conferences' held to date. POST contributes to the methodological development of TA mainly by participation in collaborations with similar offices in Europe and with academic research institutes.

Research fields

POST's work ranges across the entire field of S&T, exploring this in its economic, social and political setting. POST's structure recognises three main research areas: health and medicine, energy and environment and information technology and physical sciences. There is also an overarching concern with S&T policy. Recently, POST has introduced a particular focus into some of its work – exploring these subject areas in the context of international development.

Addressees and clients

POST's audience is mainly Parliamentarians. It frequently organises seminars and conferences linked to a project, to promote Parliamentary awareness of its output. It also monitors indicators of Parliamentary interest (e.g. members' questions) to ensure its output is targetted. However, all its work is published and it puts considerable emphasis on dissemination beyond Parliament.

Communication within the TA community

POST is an active member of the EPTA network and held its presidency in 1996 and 2002.

Contact

Parliamentary Office of Science and Technology (POST)
Westminster House
House of Commons
7 Millbank
Westminster, London SW1P 3JA
United Kingdom
http://www.parliament.uk/parliamentary_offices/post.cfm

The Institute of Technology Assessment and Systems Analysis (ITAS)

Mission

At the heart of ITAS' research work is the comprehensive analysis and evaluation of the development and application of technology and its inter-relationship with processes of societal change. Work is done on environment-related, economic and political-institutional issues and results in the development and assessment of alternative options for action and design. This is linked inexorably with the systematic reflection of normative aspects and with the further development of methods and conceptual approaches. It is an integral component of ITAS' scientific practice to communicate its results to science, politics and the general public.

Institutional setting

ITAS is a research institute within Karlsruhe Research Centre *(Forschungszentrum Karlsruhe)*. The Research Centre is member of the publicly funded *Helmholtz Association* of German Research Centres. It receives its basic funding from the Federal Ministry for Education and Research (90%) and from the German federal state of *Baden-Württemberg* (10%). ITAS is embedded into the programme-oriented structure of Helmholtz and has recently been evaluated (with high success) through an international peer review panel which applied a combined set of quality criteria consisting mainly of the familiar criteria of scientific quality as well as of criteria of societal relevance of ITAS research. Since its foundation, ITAS is running the Office of Technology Assessment at the German Bundestag (TAB, cp. XX Seitenzahl von Eintrag "TAB" einfügen).

Methodological approach

ITAS is problem-oriented with regard to its choice of subjects, organises its work in projects and is interdisciplinary with respect to scientific discipline orientation. The combination of disciplinary expertise and the ability to perform interdisciplinary analyses is essential for the projects. Beyond several methods of the disciplines involved in certain TA projects a lot of approaches and procedures from the TA toolbox play an important role for ITAS. Outstanding among these are discourse analysis, scenario technique, interview and survey techniques, risk analysis, analysis of materials flows, eco-bilancing, and practical ethics.

Research fields

The ITAS research programme is guided partly by the requirements of the institute's profile and partly by the current fields of technological advance, and their potential and risks as revealed in societal and scientific debate. Currently ITAS is active in the following research areas (1) sustainable development, (2) efficient management of resources, (3) new technologies and the information society and (4) concepts, methods and functions of problem-oriented research.

Addressees and clients

The results of ITAS' work address politics, science and industry, the societal groups affected in individual cases, and the informed general public. These results are intended to improve the information base for decision-making in politics and society and to contribute to the societal discourse. Main addressees have been in the last years several ministries at the federal and state level, national agencies and the European Commission.

Communication within TA community

ITAS makes a rich contribution to scientific communication in the field of Technology Assessment. ITAS publishes a quarterly newsletter (Technikfolgenabschätzung – Theorie und Praxis) which serves both as a platform for scientific debate and as a source of up-to-date information on national and international activities in the field of TA and on ITAS itself (see http://www.itas.fzk.de).

Contact

Institute for Technology Assessment and Systems Analysis
Research Centre Karlsruhe
P.O. box 36 40
76021 Karlsruhe
Germany
Internet: http://www.itas.fzk.de

Swiss Centre for Technology Assessment (TA-SWISS)

Mission

In a time when developments in information technology, medicine and all other fields of technology lead to changes at a rate not known before, the Centre for Technology Assessment *TA-SWISS* provides comprehensive information on the chances and risks of new technological achievements, as well as a platform for sounded and constructive dialogue.

Activities

TA-SWISS addresses technologies or technology related issues, which are considered to be controversial. In particular, it works in the three subject areas "Life Sciences and Health", "Information Society" and "Mobility". Interdisciplinary studies constitute the pillar of its activities. *TA-SWISS* studies usually view at possible technological developments and at their consequences from different angles: Questions on the impact of a technology on the environment, business, politics, human health etc. are taken up just as are those questions on the ethical implications of technical change. Besides, *TA-SWISS* implements participatory projects such as consensus conferences (PubliForum) or focus groups (publifocus). Such projects are generally built upon existing expert discussions on new technologies and aim at integrating the citizen perspective into the debate. Special procedures intended to young generations have also been developed (publitalk). Finally, *TA-SWISS* invests many efforts in communicating about its work, be it through traditional press conferences or through expert conferences or political-lunches, just to cite some examples.

Target groups

The publications and events of *TA-SWISS* are directed towards decision-makers in politics, science, business and administration, as well as towards the general public as a whole, via the media. TA-SWISS studies are expected to provide unbiased and well-balanced information in order to contribute at an early stage to putting the people and institutions who play a decisive role into the picture on the possible effects to be expected of a new technical development – and on which consequences are to be expected when this technology is not used. Participatory projects complement this strategy, in giving an independent image on the state of the social debate about technological issues.

Institutional setting

The Centre for Technology Assessment *TA-SWISS* is attached to the Swiss Science and Technology Council, which advises the Federal Council on scientific and technological issues. Its existence is stated in the federal law on research, which gives to the Swiss Science and Technology Council the mission to run prospective studies of Technology Assessment. The strategic management is in the hands of the TA-

SWISS Management Committee: it defines main areas of interest, chooses the concrete projects to work on and decides about the release of reports. Operational activities (e.g. co-ordination of studies, financial management, implementation of results, organisation of events, communication, etc.) are carried out by the *TA-SWISS* office.

Contact

Centre for Technology Assessment – TA-SWISS
Birkenweg 61
3003 Berne
Switzerland
Phone: + 41/ 31 322 99 63
Fax: + 41/ 31 323 36 59
e-mail: ta@swtr.admin.ch
internet: www.ta-swiss.ch

Center of Technology Assessment in Baden-Württemberg (CTA)

The "Center of Technology Assessment in Baden-Württemberg" is a public research foundation devoted to the systematic and non-partisan study of technical, social, and economic impacts of technological change. Its main goal is to generate, collect and promote better knowledge on how different actors in economics, politics and society can contribute to a humane and sustainable future with respect to techno-logical and social development. The foundation was established by the German State of Baden-Württemberg in 1992. It employs approximately 70 staff members from all major disciplines. The main characteristics of the Center are:

– to conduct interdisciplinary and transsectoral research on the likely consequences of technological change on the economy, society and the individual with a particular focus on the German State of Baden-Württemberg;
– to provide unbiased, independent and professional expertise about the determinants and impacts of technological change;
– to develop a network of TA research institutions in Baden-Württemberg and beyond
– to consult policy and decision makers with respect to technological choices and potential consequences of different policy options.
– to involve all relevant and affected social groups about technological impacts in discursive or participatory activities.

The Center is structured in four scientific departments, the department "administration and public relations" and a horizontal unit on discourse activities. The main fields of research and policy consulting are: regional concepts of sustainable development, innovations for economic advancement, labor relations and employment, impacts of technical and social infrastructure on quality of life, implications and impacts of life sciences, methods and procedures of public discourse (horizontal activity). Within each subject area, a team of investigators from the technical, natural, and social sciences pursues research projects aimed at assessing and evaluating im-pacts of specific technologies within a broader subject area such as energy or sustainable development. While the assessment procedure is mainly dominated by scientific processes, interpretation and evaluation of results are subject to an extensive discourse among researchers, politicians, representatives of social groups, and the public at large. The involvement of relevant actors and affected citizens is part of the legal mandate of the Center and one of its major character-istics. The Center has experienced with many innovative forms of stakeholder involvement and public participation on a multitude of issues ranging from resolving siting conflicts to collecting public input for regulating biotechnological risks.

The Center is not affiliated to any branch of government nor related to any private interest. This status of foundation grants the Center scientific and political independence and provides it with the financial and institutional strength to undertake studies for third parties (as long as they are in the public interest). In the past, the Center received research grants from political organizations (such as Communities, German States or the Federal Government, the EU, the OECD, the WHO and others), research institutions and agencies (such as the German Research Foundation) and private organizations and foundations. Due to the Center's professional and flexible administration, a dedicated staff of researchers from all relevant fields of the sciences and the humanities, a product-oriented style of work and a 10 year experience with external research grants the Center offers ideal conditions for all parties interested in genuine and reflective research on technological change, communication strategies and public involvement.

Contact

Center of Technology Assessment in Baden-Württemberg (CTA)
Industriestraße 5
70565 Stuttgart
Germany
(comment: CTA has been closed down on the 31.12.2003)

The Danish Board of Technology (DBT) – Teknologirådet

Mission

The DBT initiates comprehensive technology assessment activities on the opportunities and consequences of technology for the society and for the individual citizen. The Board should initiate independent assessments, and communicate the results of its work to the Parliament, to other political decision-makers, and to the Danish people in order to support and enhance the public debate on technological issues. It is advisor for the Parliament as well as for the government.

Institutional setting

The DBT is established by the Parliament as an independent public institution, connected to the Ministry of Science. The minister appoints the Board of Governors (BG) and the Board of Representatives (BR). The BG has 11 members of whom 7 are nominated by institutions from the labour market, innovation system and research councils, local governments and public education. The BG is responsible for the budget, the work programme, and the establishment of the secretariat. The BR has 50 members, nominated by a broad scope of institutions, and function as a network for discussion of the work of the Board. The secretariat has 8 academic and 4.5 administrative staff man-year, plus project internals (students, consultants), trainees, social programme employees etc. The budget of 2003 was 10.3 mill.DKK ~ 1.4 mill.€.

Methodological approach

DBT makes use of five clusters of methods: a) expert-based studies, b) stakeholder involvement, c) citizen consultation, d) public communication, and c) political advisory methods. The BG makes the work programme after an open call for ideas, involving call letters, a workshop and an open internet scheme. The 1–200 ideas are rated on six criteria and prioritised. 20 top candidate ideas are researched further to support the final choice of work-plan. As part of the research the problem situation is described, the main problems selected, and a project design sketched. This procedure reflects the methodological principles of the DBT that the method should be chosen as a best possible answer to the problem situation and the processes should be as transparent as possible. The project should ideally be able to include the necessary knowledge, make normative clarifications possible and nourish the ground for decisions and change.

Research fields

In practise, any societal problem involving technology or in fields that are closely engaged in scientific and technological developments can be taken up. Such fields are for example: Energy, environment, work environment, biotech, food, transportation, ICT, health care, housing and planning, intellectual property rights.

Addressees and clients

Depending on the issue, the DBT involves or communicates results to politicians (parliament, government, counties, municipalities, EU level), governmental institutions, science and the innovation systems, the labour market actors, NGO's and other stakeholders, the education system, and the public. The target groups can be broad or narrow, also depending on the issue. The DBT sees the media as a gatekeeper that needs special attention, and a communication officer is connected to all project activities to enhance the media impact.

Communication within TA community

The DBT is actively involved in international cooperation on TA and Foresight. It is continuously involved in methodological research, consultancy activities, networks, EU projects, conference presentations, international visits etc. Especially in the field of public participation, the DBT has distributed experience and methodology capacity both inside and outside the TA community. DBT has an informative English website at www.tekno.dk.

Contact

Danish Board of Technology (DBT)
Antonigade 4
1106 Copenhagen
Denmark
Phone: +45/33 32 05 03
Fax: +45/33 91 05 09
Email: tekno@tekno.dk
www.tekno.dk

Office of Technology Assessment at the German Parliament (TAB)

TAB's primary task is to design and implement technology assessment (TA) projects. These projects are supplemented by projects analysing important scientific and technological trends and the associated social developments (monitoring). TAB is expanding its range of activities by contributing to long-term technology foresight, and analysing international policies and innovation developments. As seen by TAB the objectives of TA as an advisory function for policy are,

- to analyse the potentials of new scientific and technological developments and identify and explore the associated social, economic and ecological opportunities,
- to examine the legal, economic and social framework conditions for implementing scientific and technological developments,
- to provide a comprehensive analysis of the potential impact of future utilisation of new scientific and technological developments and indicate the possibilities for strategic exploitation of the potential uses of technologies and avoiding or reducing the associated risks and from this basis
- to develop alternative options for action and guidance for political decision makers.

Mandate

TAB is an independent scientific institution created with the objective of advising the German Bundestag and its committees on matters relating to research and technology.

TAB is operated by the Institute for Technology Assessment and Systems Analysis (ITAS) at the Karlsruhe Research Centre and is an operational unit of ITAS. In executing its working programme the Karlsruhe Research Centre cooperates with the Fraunhofer Institute for Systems and Innovation Research (ISI), Karlsruhe, starting from September 2003. TAB's work is strictly oriented towards the information needs of the Parliament and its committees. The political control organ is the Committee on Education, Research and Technology Assessment, which decides on the initiation of TA projects. Motions to start a TA project can be submitted by parliamentary political groups on the Committee on Education, Research and Technology Assessment or in other committees. TAB's director has scientific responsibility for the working results of TAB. Rapporteurs from one or more committees assist TAB projects and help integrate the results into the committees' activities. The findings of TA projects and other TAB activities are primarily made available in the form of TAB working reports. The final reports on TA projects are also published as printed papers of the German Parliament (Bundestagsdrucksache). Since 1996 final reports on TA projects have appeared in a series of books ("Studies by the Office of Technology Assessment at the German Bundestag").

Activities 2003

(for detailed information see: www.tab.fzk.de)

- Trends in food supply and demand and their implications
- Reduction of acreage utilisation
- Analysis of network-based communication with regard to cultural aspects
- Green genetic engineering
- Modern agrarian technologies and production methods
- Preimplantation diagnostics
- New forms of dialogue between science, politics and the public
- Biometric identification systems
- Lighter-than-air aircraft technology
- eLearning
- Demand-oriented innovation policy
- The future of gainful occupation – working in the future

Contact

Büro für Technikfolgen-Abschätzung beim Deutschen Bundestag (TAB)
Neue Schönhauser Str. 10
10178 Berlin
Germany
Phone: +49/30 28491-0
Fax: +49/30 28491-119
email: buero@tab.fzk.de
http://www.tab.fzk.de/

The European Parliament (EP)

The European Parliament is a direct descendant of the ECSC (European Coal and Steel Community) Common Assembly and the "European Parliamentary Assembly" that was created following the entry into force of the EEC (European Economic Community) and Euratom Treaties in 1958 and changed its name to "European Parliament" on 30 March 1962. Initially the Members of the European Parliament (MEPs) were appointed by each of the Member States' national parliaments. The first direct European elections were held in June 1979 and the sixth will be held in June 2004. Following successive enlargements of the European Union the Parliament currently has 626 members from 15 Member States. This number will further increase with the accession of 10 new Member States in May 2004. After an original extension of its budgetary powers in 1970, the Parliament saw its legislative powers increase, first (cooperation procedure) with the entry into force of the Single European Act on 1 July 1987 and later (codecision procedure, approval of new Commission) with the entry into force of the Maastricht Treaty on 1 November 1993. Following the entry into force of the Amsterdam and Nice Treaties (1 May 1999 and 1 February 2003, respectively) the codecision procedure was extended to most areas of legislation and Parliament emerged as co-legislator on an equal footing with the Council. The preliminary but substantial stages of legislative work are carried out in a number of parliamentary committees and all members of the Parliament are assigned to such committees (including subcommittees and occasional temporary committees). In the sixth legislature (2004–2009) the European Parliament will have twenty standing committees. One of them will be the Committee on Industry, Research and Energy. According to a resolution adopted by the plenary on 29 January 2004, its competences will be as follows:

– the Union's industrial policy and the application of new technologies, including measures relating to SMEs;
– the Union's research policy, including the dissemination and exploitation of research findings;
– space policy;
– the activities of the Joint Research Centre and the Central Office for Nuclear Measurements, as well as JET, ITER and other projects in the same area;
– Community measures relating to energy policy in general, the security of energy supply and energy efficiency including the establishment and development of trans-European networks in the energy infrastructure sector;
– the Euratom Treaty and Euratom Supply Agency; nuclear safety, decommissioning and waste disposal in the nuclear sector;
– the information society and information technology, including the establishment and development of trans-European networks in the telecommunication infrastructure sector.

Technology assessment (TA) work within the European Parliament falls within the remit of STOA (Scientific and Technological Options Assessment), whose purpose is to: "to assist Parliament's committees, at their request, in the performance of their legislative tasks and their activities of a general nature in which science and technology play a predominant role" (EP Bureau decisions of 18 September 1995 and 17 February 1997). On 10 October 1985 the European Parliament adopted the report by Mr Rolf LINKOHR, "on the establishment of a European Parliament Office for Scientific and Technological Option Assessment". The report stresses "the particular needs of the standing committees and political groups in technical and political decision-making, which can be met only by an autonomous technology assessment office". It then goes on to propose "that a European Parliament office for scientific and technological option assessment should be set up ... to coordinate assessment work and award external contracts in support of its work". The now defunct Office of Technology Assessment (OTA) of the US Congress served at that time as a model.

Following a decision of the Parliament's Bureau in June 1986, STOA was officially launched in March 1987, as an 18-month pilot project under the responsibility of the then Committee on Research, Technological Development and Energy. In September 1988 a report was submitted to the Bureau on STOA's performance during the pilot period. As a result, the Bureau authorised STOA to continue its work on a permanent basis, on condition that it would make its services available to all standing parliamentary committees. The STOA Panel was created and all standing committees were invited to nominate their representatives to the Panel.

STOA was transferred from the Directorate-General II (DG II, Committees and Delegations) to the Directorate-General for Research (DG IV) in 1991 and its secretariat was fully integrated in the relevant Division of the latter in 2000. The external research budget of DG IV had been divided equally between the general studies programme and STOA until the end of 2003. In a note on the "future mandate, activities and organisation" of STOA submitted by the EP Secretary General to the Parliament's Bureau on 25 August 2003, it is made clear that "[t]he reasons which led Parliament to set up STOA, as summarised in the Linkohr Report in 1985 remain at least as valid today as when the project was first mooted". On 1 September 2003, the Bureau approved the Secretary General's proposal for a reform of STOA, deciding among, other things, "to organise the political oversight of STOA, on the basis of a small panel of members to be responsible for approving the programme ...". According to the proposal, "[t]he panel could be composed of seven members ...", "... essentially drawn from the committee with responsibility for research". The activities of STOA should include medium- and long-term studies on scientific and technological developments, carried out by an external high-quality scientific network, forums and panels, as well as cooperation with similar national organisations, bilaterally and through the European Parliamentary TA (EPTA) network, and a developing partnership with the European Academies' Science Advisory Council (EASAC).

Contact

European Parliament
Committee on Industry, External Trade, Research and Energy
Rue Wiertz 60 – ATR 06K066
1047 Brussels
Belgium
http://www.europarl.eu.int/

Centre for Science, Technology, Society Studies at the Institute of Philosophy, Academy of Sciences (STS Centre)

The Centre conducts systematic interdisciplinary research into mutual relationships of science, technology, and society. Research is based on the broad theoretical background of philosophy, economy, and sociology (of science and technology), innovation studies, science policy studies, and technology assessment. At present, research is focused on several fundamental thematic areas. The changes in the current forms of knowledge production are followed in the form of theoretical analysis. Those changes comprises higher rate of social accountability of knowledge, more emphasis on its economic application, increasing heterogeneity of its resources, the problem of evaluation, more stress on social and cultural values applied in research programmes. The emerging knowledge society is being analysed from the viewpoint of transformation of science and research. Both theoretical and practical issues are analysed in the fields of science policy, innovation policy, research evaluation, technology assessment. The key following problems are under investigation: research as source of innovations, integration of research into national innovation system, integration of national innovation system into international one, new ways of bridging the gap between academe and industry in knowledge society.

The Centre deals also with more practically oriented problems namely in the area of innovation studies. It has been conducting several empirical surveys focused on innovation level of the Czech R&D institutions and enterprises. Also some case studies analysing the innovative situation in Czech industry were completed. The Centre is involved into elaborating benchmark of the Czech R&D institutions – namely through the participation in RECORD network. The Centre draws on the work that was done during the 90ies in analysing the transformation process of the Czech research system. Then, it was gained relevant information on the structure and orientation of "transformed" research institutions. Participation in international projects:

- 1993–98 Network on Transformation of Science Systems in the Central- and East European Countries, German Federal Ministry of Research
- 1999–2000 IPTS Futures Project
- 2001–2003 ESTO – European Science and Technology Observatory (FP5)
- 2002–2003 TAMI – Technology Assessment in Europe: Between Method and Impact)
- 2002–2004 RECORD – Recognising Central and Eastern European Centres of R&D: Perspectives for the ERA, STRATA Programme of FP 5.

Contact

Centre of Science, Technology, Society Studies at the Institute of Philosophy
Academy of Sciences of the Czech Republic (STS Centre)
Jilska 1
11000 Prague
Czech Republic
email: stsscz@cesnet.cz

The Institute of Modern Civilization (SGH)

The Institute of Modern Civilization is an interuniversity unit that acts on the basis of an agreement signed in 1996 by the Medical University of Warsaw, Warsaw University, Warsaw School of Economics and Warsaw University of Technology. Supervision over the Institute is carried out by the Council of the Institute, consisting of Rectors of the Founding Universities and Secretary or Under-Secretary of State in Ministry of National Education. The Institute carries out studies on social understanding of modern civilization, predicts influence of the science and technology development on the future shape of society, especially academic society, undertakes actions aimed at universalization of social consciousness, acts in favor of development and integration of academic society.

Scientific staff of the Institute do research work, take educational and organizational actions in favor of an academic society in the whole Poland, but especially in Warsaw. At the Institute several series of seminars are organized. One also elaborates analyses and research projects. The results of these works are presented, among others, in a series of reports published by the Institute. The Centre for Information Society, a unit within the Institute of Modern Civilization, organizes academic courses on different aspects of information society for master and doctor degree students of Warsaw universities.

Origin and mission of the Institute

In its activities the Institute of Modern Civilisation combines global and local aspects of the modern society and is especially interested in practical applications of the science. Research carried out by the Institute focus on the analysis of the challenges arising from the civilization and cultural changes, faced by the society that enters the third century. Seminars organized by the Institute are accompanied by advisory activities and lead to a number of studies, research projects, articles and seminar proceedings. Established by Warsaw universities, the Institute intends to act for integration of the Warsaw academic society. At the same time it remains open to all societies and during its lectures and seminars offers interdisciplinary working basis for exchange of opinions on general subjects.

Research projects and studies

Most important projects completed in the Institute are:

- Academic Accreditation Committee – the System of Assessment of Training Quality and Accreditation in Higher Education, by order of the Ministry of National Education;
- Poland Facing European Integration in Education – Proposals and Threats, by order of the Committee of European Integration;
- Interuniversity Education, by order of Founding Universities;
- The Cost Analysis in Higher Education, by order of the Ministry of National Education.

Lectures

During every academic year the Centre for Information Society within the Institute organizes a series of courses for Warsaw students. Among the lecturers there are professors from Medical Academy of Warsaw, Warsaw University of Technology, Warsaw Academy of Arts, Warsaw School of Economics, University of Warsaw and Institutes. At the end of every course students write examination that is recognized by the universities and confirmed with a special certificate.

Contact

Warsaw School of Economics – Institute of Modern Civilisation (SGH)
Division of Decision Analysis and Support
Niepodleglosci 164
Room 614
02-554 Warsaw
Poland
http://www.ipwc.pw.edu.pl/

CSIC Unit on Comparative Policy and Politics

Spanish Policy Research on Innovation & Technology, Training & Education (SPRITTE) Group

Consejo Superior de Investigaciones Científicas (CSIC)

Unidad de Políticas Comparadas (UPC)

The Spanish National Scientific Research Centre (CSIC, Consejo Superior de Investigaciones Científicas), established in 1939, is an overarching interdisciplinary organisation that carry out research through its one hundred institutes in all disciplines and research fields. The CSIC, the largest Spanish research institution, and more than ten thousand people work all over the country.

The CSIC has diverse institutes, centres and units dealing with social sciences and policy analysis; One of them is the CSIC Unit on Comparative Policy and Politics, specialised in producing research and analysis of the policy and political processes, in regions like Europe and Latin America and in areas like social policies, education and science and technology policies. The research group working in the domain of science and technology policies is SPRITTE (Spanish Policy Research on Innovation & Technology, Training & Education) under the leadership of Dr. Luis Sanz-Menéndez.

The objective of SPRITTE is to increase knowledge in the study of science, technology, innovation, training and education systems, the strategies of the actors within them, using a comparative international perspective, and emphasizing the role public policy plays in these environments. SPRITTE is formed mainly by sociologists and political scientists. The aim of SPRITTE is to contribute to the increase of knowledge in the field of science, technology and innovation studies, through the analysis of the structures and dynamics of change of research and innovation systems and through the understanding of the behaviour of research and innovation actors (firms, laboratories, institutions, governments etc.) and to the explanation of the driving forces in S&T and innovation policies from a comparative perspective. The SPRITTE Group seeks to better comprehend the growing importance of the denominated knowledge economy and society of this upcoming century.

There is also at CSIC UPC a research group on "science, technology and society" (STS), that has been working since 1991 under the leadership of Prof. Emilio Muñoz on the development of studies field with a double orientation: i) applying a diversity of methods and strategies to these studies (based on history, sociology and evaluation approaches); ii) selecting a series of scientific disciplines and technologies, related to biology and biomedicine (biotechnology, biochemistry and molecular biology, biomedicine, fish farming and food). This strategy has consolidated in a number of research lines such as: evaluation of science and technology policies; changes in the production of knowledge regimes; ethical issues related to technology (genetic modified food, xenotransplantation, steam cells, etc.); gender issues in science and technology; the emergence and evolution of modern biology in Spain and the science policy in the 20th century; biomedical experiences in the relation between science and industry in Spain.

International research collaboration is developed under a permanent basis mainly, through research projects, with some European research institutes such us: PREST University of Manchester, CSI from Paris, Fraunhofer Institute for Systems. and Innovation Research (ISI-FhG Karlsruhe), Twente University, SPRU, etc.

Contact

CONSEJO SUPERIOR DE INVESTIGACIONES CIENTÍFICAS (CSIC)
CSIC Unit on Comparative Policy and Politics
SPRITTE- Spanish Policy research on Innovation & Technology, Training & Education
C/ Alfonso XII, n. 18, E-28014 Madrid, Spain
http://www.iesam.csic.es/

The Rathenau Institute

Mission

The Rathenau Institute's task is to support societal debate and political opinion forming concerning issues resulting from or linked to scientific and technological developments, including ethical questions.

Institutional setting

The Rathenau Institute was created by ministerial decree of June 17th 1986 as an independent governmental organisation. It is one of the institutes of the Royal Netherlands Academy of Arts and Science (KNAW) that is located in Amsterdam. The KNAW, and accordingly the Rathenau Institute, receives its basic funding from the Ministry of Education, Culture and Science.

Methodological approach

To improve the quality of the political and public debate on societal aspects of science and technology, the Rathenau Institute uses a colourful set of instruments, like questionnaires, citizen panels, scenarios, movies, hearings, fiction, et cetera. Depending on the perceived shortcomings of the debate one or a mix of instruments are employed. Over the years the Institute has broadened its methodological horizon. Starting off as research-oriented organisation, the Rathenau Institute began in the beginning of the 1990s to experiment with participatory TA, including public participation through public panels and interactive TA methods. Over the last years the method toolbox has been broadened by the use of various art forms, like theatre, movies, and role-plays. These types of methods are used to visualise possible futures, and thereby to enable a more concrete discussion about those futures. Such instruments not only inform, but also entertain. The two-yearly Rathenau Technology Festival also falls under this new heading of infotainment. During this event various forms of debate, film, expositions, and cabaret come together.

Research fields

The Rathenau Institute has a two-year working programme. The selection of projects is guided by three criteria. Subjects must contain a technological and scientific component. Subjects must be socially, politically and/or morally relevant. And social or political debate must be needed or desired. Besides the Institute strives after a balance between new and existing technology, and topical issues and subjects that are not (yet) on the political agenda.

The current working programme 2003–2004 deals with the following broad range of topics: organ replacement, reproductive medicine, everyday medical technology, influencing behaviour through technology, animal welfare, vulnerability of the information society, food genomics, healthy eating, nanotechnology, military technology, renewed nuclear energy debate, anti-ageing technology, intelligent interaction between man and machine, cyborgs, reproductive cloning of humans, and (top-class) sport.

Addressees and clients

In our projects a wide range of actors – citizens, stakeholders, and experts – are involved. Project results are communicated towards a broad interested audience, but in particular are addressed to parliament, as a representative of the Dutch people and as a central place of democratic decision making.

Contact

Rathenau Institute
Koninginnegracht 56
2514 AE Den Haag
Postbox 85525
2508 CE Den Haag
The Netherlands
http://www.rathenau.nl

Flemish Institute for Science and Technology Assessment (viWTA)

Mission

The 'Vlaams Instituut voor Wetenschappelijk en Technologisch Aspectenonder-zoek' was founded by decree on July 17, 2000 by the Flemish Parliament. Its mission was defined as: 'to study the aspects and consequences of developments in science and technology, focusing on their impact on society'. viWTA will inform the Flemish Parliament about public debate and controversies involving science and technology.

The main role of viWTA is thus advicing the Flemish Parliament, either on its own initiative or in response to questions from Members of the Flemish Parliament. Through its advizes, viWTA wants to contribute to a transparent debate about complex issues involving society and technology. In the public debate, it is the role of viWTA to clarify arguments and positions, to interpret the subjects in their context, to elucidate the debate and to see to it that in addition to experts the general public can also be heard.

Institutional setting

As an autonomous parliamentary institution, viWTA has its own Board of Directors. The Board consists of sixteen members, half of whom are Members of the Flemish Parliament (from whom the president is elected), while the other half is composed of experts from the Flemish scientific, ecological and socio-economic community. ViWTA receives his complete annual working budget from the Flemisch Parliament

Methodological approach

Without neglecting traditional methods of TA like experts surveys and risk analysis, the viWTA is very devoted to participative and communication methods. All too often discussions about science and technology are restricted to a small group of experts and researchers. One of the aims of viWTA is to allow the opinion of the general public to be heard in the debate about the possibilities and limitations of new technologies. By providing comprehensive and balanced information concerning evolutions and opinions, viWTA wishes to involve the general public in the debate about science and technology in a systematic and scientifically sound way.

After all, it is by combining scientific research and public involvement, that the Flemish Parliament will have the opportunity to keep their policy-making processes in touch with society. This broad approach is reflected in the composition of the scientific secretariat of the institution. Its staff consists of the Director, the administrative secretariat, a group of researchers who co-ordinate the research and supervise the participative and interactive initiatives and a communications officer.

Working program

The research and the activities within the viWTA are conducted upon the basis of a yearly working program. The ideas originate from three main sources. Firstly, (members of) the Flemisch Parliament can ask questions or put forward themes for research. Secondly, everybody can post suggestions for new projects (on our website, by e-mail or a letter). Thirdly, also the staff can generate new ideas, based on regularly undertaken monitoring exercises. Besides this working program, it is always possible that the viWTA gets (short-time) questions from commissions or members of the parliament. In his – still very young – history, the main fields of research and activity were GMO's, energy and climate and projects regarding theory and practice of TA.

Communication

Five times a year, viWTA publishes a newsletter with information on his projects, small flashes of interesting national or international findings, an agenda of activities,... The results of the research projects or the organised activities are mostly published in a report or – if not – a smaller publication. All produced material is also available via the website: www.viwta.be.

Contact

viWTA-Samenleving en Technologie
Vlaams Parlement
1011 Brussels, Belgium
http://www.viwta.be./

List of Authors

Bellucci, Sergio, Dr, is Director of the Swiss Centre for Technology Assessment (TA-SWISS) at the Swiss Science and Technology Council. He occupies this position since 1996. He received a degree in agronomy of the Swiss Federal Institute of Technology in Zurich in 1976, and a doctorate at the same Institute in 1980. He then became researcher at the Agro-Division of Ciba-Geigy in Basle. In 1989, he became Director of the Centre for Continuing Education of the Swiss Federal Institute of Technology in Zurich and, in 1993, Director of the Management and Technology Institute at the Technopark in Zurich. Dr. Sergio Bellucci is member of several groups, commissions and projects addressing the issue of "science, technology and society", such as the European Parliamentary Technology Assessment Network EPTA and the International Network of Agencies for Health Technology Assessment INAHTA.

Postal address: Centre for Technology Assessment at the Swiss Science and Technology Council (TA-SWISS), Birkenweg 61, 3003 Bern, Switzerland.

Email: Sergio.Bellucci@swtr.admin.ch

Berloznik, Robby, studied political science and philosophy at the Free University of Brussels (1979). After a research career in technology assessment he entered the Flemish Institute for Technological Research in Mol in 1991 were he became the advisor for technology assessment to the managing director. This until 1997 when he became research manager in the fields of technology assessment, technology foresight and sustainable development. In December 2001 he was appointed by the Flemish Parliament as the first director of the Flemish Institute for Science and Technology Assessment (viWTA). This institute supports the Flemish Parliament in its science and technology decision making.

Postal address: viWTA-Samenleving en Technologie, Vlaams Parlement, 1011 Brussels, Belgium.

Email: robby.berloznik@vlaamsparlement.be

Bütschi, Danielle, Dr, studied political science at the University of Geneva. She received a degree from the Social Sciences and Economics Faculty in 1989, and a doctorate from the same Faculty in 1997. From 1991 to 1996, she conducted researches at the University of Geneva (Department of Political Science) on NGOs role in welfare policies and on public opinion related to environmental issues. From 1997 to 2001, she worked at the Swiss Centre for Technology Assessment (TA-SWISS) as project manager. She was in charge of participatory procedures and of Information society projects. Since 2002, she works as an external partner of TA-SWISS, as well as for other institutions. She is mainly active in projects based on a dialogue between science and society.

Postal address: Centre for Technology Assessment at the Swiss Science and Technology Council (TA-SWISS), Birkenweg 61, 3003 Bern, Switzerland.

Email: Danielle.Buetschi@swtr.admin.ch

Carius, Rainer, Master of Science in Mechanical Engineering, University of Darmstadt and Ecole Central de Lyon, Master of Science in Management, Troy State University Alabama. Rainer Carius is a senior researcher in the department for technology, society and environmental economics and Assistent Head of Discourse Departement at the Center of Technology Assessment in Baden-Württemberg. He served as head of several research projects concerning regional concepts of sustainable development, shareholder and public participation, standards of environmental quality and risk communication and management. Rainer Carius is co-author of two books and has published about 30 articles. He also lectures at the Universities of Applied Sciences in Mannheim and serves on different advisory boards on a regional and national level.

Postal address: Center of Technology Assessment in Baden-Württemberg, Industriestr. 5, 70565 Stuttgart (until 31.8.2003), Germany.

Email: rainer.carius@uvm.bwl.de (from 1.9.2003)

Cope, David, Professor, has been Director of the Parliamentary Office of Science and Technology at the Houses of Parliament since 1998. He returned to the UK to take up this position from Kyoto in Japan, where he was Professor of Energy and Resource Economics. For 11 years before that, he had been Executive Director of a charitable foundation research centre in Cambridge, while earlier in his career he has been Environmental Team Leader with the International Energy Agency and Lecturer in Interdisciplinary Studies at the University of Nottingham. He has held visiting fellowships at the University of Hong Kong and the East-West Center, University of Hawaii and won a scholarship from the US Embassy, London, to study Technology Assessment in the US in 1979.

Postal address: Parliamentary Office of Science and Technology, Westminster House, House of Commons, 7 Millbank, Westminster, London SW1P 3JA, United Kingdom.

Email: COPED@parliament.uk

Cruz-Castro, Laura, Dr, is Research Fellow at the Spanish Higher Council for Scientific Research (CSIC). She is BA in Political Science and Sociology, Master in Social Sciences, and PhD in Sociology, Autonomous University of Madrid. She has been Visiting Researcher at Nuffield college (Oxford). Her research area is science and technology policies and systems. She has been involved in professional international research conferences and European Commission experts' activities. She has published in national and international journals in issues related to education, science and research centres.

Postal address: Consejo Superior de Investigaciones Científicas (CSIC), Unit on Comparative Policy and Politics, C/ Alfonso XII, n. 18, 28014 Madrid, Spain.

Email: Laura.Cruz@iesam.csic.es

Decker, Michael, Dr rer. nat., studied physics (minor subject economics) at the university of Heidelberg, 1992 diploma, 1995 doctorate with a dissertation on temperature measurements in high pressure combustion by laser-induced fluorescence of molecular oxygen at the university of Heidelberg, 1995–1997 scientist at the German Aerospace Center (DLR) in Stuttgart, 1997–2002 member of the scientific staff of the Europäische Akademie GmbH. He was manager of the project "Robotics. Options of the replaceability of human beings" and manages the project "Technology Assessment between Method and Impact" and the study group "Miniaturization and Material Properties". Since 2003 he is member of the scientific staff and since February 2004 vice-director of the Institute for Technology Assessment and System Analysis (ITAS) at the Research Centre Karlsruhe. Main research areas: TA of robotics and nanotechnology, comparison of TA-methods and interdisciplinary research.

Postal address: Institute for Technology Assessment and Systems Analysis, Research Centre Karlsruhe, P.O. box 36 40, 76021 Karlsruhe, Germany.

Email: Michael.Decker@itas.fzk.de

Gram, Søren, Dr, is Project Manager at the Danish Board of Technology. He occupies this position since 1997. Current fields of responsibility are Technology in developing countries, globalization, biotechnology, climate changes, electronic surveillance, public health care, city traffic, co-operation between science and business community, and dialogue methods to involve citizens. He received a degree in Physics of the University of Copenhagen in 1980, and a degree in Cultural Sociology at the same University in 1988. In 1997 he received a PhD in Natural Resource Management of the Royal Danish Veterinary and Agricultural University. From 1989 to 1992 he worked as a Consultant at the Centre for Alternative Social Analyses. From 1992 to 1996 he worked as a Researcher at the Royal Danish Veterinary and Agricultural University. And from 1996 to 1997 he worked as a Research fellow at the Centre for Development Research.

Postal address: Danish Board of Technology (DBT), Antonigade 4, 1106 Copenhagen, Denmark.

Email: sg@Tekno.dk

Grunwald, Armin, Professor Dr rer. nat., studied physics at the universities Münster and Cologne, 1984 diploma, 1987 dissertation on thermal transport processes in semiconductors at Cologne university, 1987–1991 software engineering and systems specialist, studies of mathematics and philosophy at Cologne university, 1992 graduate (Staatsexamen), 1991–1995 scientist at the DLR (German Aerospace Center) in the field of technology assessment, 1996 vice director of the Europäische Akademie GmbH, 1998 habilitation at the faculty of social sciences and philosophy at Marburg university with a study on culturalistic planning theory. Since October 1999 director of the institute for Technology Assessment and systems analysis (ITAS) at the research center Karlsruhe and professor at the university of Freiburg. Since 2002 also director of the Office of Technology Assessment at the German Bundestag (TAB). Since 2003, in addition, speaker of the Helmholtz programme "Sustainable Development and Technology". Working areas: theory and methodology of technology assessment, ethics of technology, philosophy of science, approaches to sustainable development.

Postal address: Institute for Technology Assessment and Systems Analysis, Research Centre Karlsruhe, P.O. box 36 40, 76021 Karlsruhe, Germany.

Email: Armin.Grunwald@itas.fzk.de

Hennen, Leonhard, Dr phil., Sociologist M.A., studied sociology and political sciences. After five years as a researcher at the department of "Technology and Society" at the National Research Center, Jülich (projects on "technology and everyday life", "risk-communication"), he is since 1991 member of the scientific staff of the Office of Technology Assessment at the German Parliament. Among others he has been responsible for projects on Genetic Testing, Technology controversies, Sustainable Development and Research Policy. He is responsible for the Office of TAs international contacts: e.g. to the European Parliamentary Technology Assessment Network (EPTA).

Main research interests: sociology of technology, technology policy, concepts and methods of technology assessment.

Postal address: Büro für Technikfolgenabschätzung beim Deutschen Bundestag

Neue Schönhauser Str. 10, 10178 Berlin, Germany.

Email: hennen@tab.fzk.de

Karapiperis, Theodoros, Dr, was born and went to primary and secondary school in Thessaloniki Greece. After a year in Sevenoaks School, UK, he read Mathematical Physics at the University of Sussex, UK, and went on to earn a PhD in Theoretical Nuclear Physics from the Massachusetts Institute of Technology in 1982. He worked for several years in fundamental research (Schweizerisches Institut für Nuklearforschung, Villigen, Switzerland, and Universität Erlangen-Nürnberg, Erlangen, Germany) before switching in 1990 to applied research (Group Modezeitung Endlager-Sicherheit at the

Paul Scherrer Institut, Villigen and Würenlingen, Switzerland). In 1995 he took up a post as administrator at the European Parliament in Brussels. After a stay at the Press Service of the Parliament, he moved to the secretariat of STOA (Scientific and Technological Options Assessment) and the secretariat of the Committee on Industry, External Trade, Research and Energy, where he deals mostly with matters related to Community research policy, including all aspects of the Research Framework Programme.

Postal address: European Parliament, Secretariat of the Committee on Industry, External Trade, Research and Energy, Rue Wiertz 60 – ATR 06K066, 1047 Brussels, Belgium.

Email: tkarapiperis@europarl.eu.int

Klüver, Lars, director of The Danish Board of Technology – Teknologirådet, which is the Danish parliamentary technology assessment institution. MSc in ecology and environmental biology, University of Copenhagen, 1984. From 1982–1989 Lars Klüver had a private consultancy enterprise on communication and audiovisuals. Project manager at The Danish Board of Technology as a project manager 1986–1990; head of project unit 1990–1994, and director since 1994. Lars Klüver has been involved in numerous international conferences and projects on Technology Assessment and its methodology. He coordinated the EUROPTA project (European Participatory Technology Assessment), was a member of the EU Commission's High Level Expert Group on Foresight, and has been involved in many advisory and evaluation activities on TA and foresight programmes and institutions.

Postal address: Danish Board of Technology (DBT), Antonigade 4, 1106 Copenhagen, Denmark.

Email: LK@Tekno.dk

Ladikas, Miltos, Dr, (nee. Liakopoulos) studied Psychology in Athens (BA, Deree College) and Social Psychology in London (MSc, PhD, London School of Economics) with a thesis analysing public debates of Life Sciences in Europe; since 1994 he has held various research positions at the London School of Economics, UK (socio-economic implications of Life Sciences) and the Science Museum in London, UK (public perceptions of Science & Technology). He has been member of the coordination team of the EC Concerted Actions "Ethical Dimensions of Biotechnology: The Bases of Public Concerns" and "Biotechnology and the European Public". Since 2000 he is member of the Europäische Akademie GmbH, as coordinator of the project "Functional Foods" and also co-coordinator of the EC-funded project TAMI (Technology Assessment in Europe: between Method and Impact). His publications and research interests are in the areas of ethics in technology assessment, public perceptions of S&T, food technologies, modern biotechnologies and science & technology policy.

Postal address: Europäische Akademie GmbH, Wilhelmstr. 56, Bad Neuenahr-Ahrweiler, Germany.

Email: m.ladikas@lancaster.ac.uk

Machleidt, Petr, Dr phil., (born 1949), studied sociology and economy at the Faculty of Philosophy, Charles University in Prague, 1974 diploma, 1976 doctorate at the Charles University. From 1974 member of research staff (in Academy of Sciences) in the field of science, technology, society relations, 1975–1990 at the Institute of Philosophy and Sociology of the former Czechoslovak Academy of Sciences. Currently (since 1990) member of the Institute of Philosophy of the Academy of Sciences, Czech Republic. Main research areas: social and human dimensions of science and research assessment, Transformation of science systems, Technology Assessment, Comparison of TA-Concepts. Member of international network on transformation of research systems. Co-autor of several journal articles and contributions to the books concentrating on transformation of the Czech republic.

Postal address: Centre of Science, Technology, Society Studies at the Institute of Philosophy, Academy of Sciences of the Czech Republic (STS Centre), Jilska 1, 11000 Prague, Czech Republic.

Email: stsscz@cesnet.cz

Sanz-Menéndez, Luis, Dr, is Senior Research Scientist of the Spanish Higher Council for Scientific Research (CSIC). Before joining the CSIC he was associate professor at the Political Science and Sociology School of the Complutense University of Madrid. And from 2000 to 2002 he was Deputy Director General for Research Planning and Monitoring at the Spanish Ministry of Science and Technology. He has more than 20 years of research experience in the European and Latinamerican science and technology systems and he also has been advising national and international organizations such as the OECD, European Union, UNESCO and UNIDO. He has more than seventy publications and works on science, technology and innovation policies in international journals like Research Policy, Research Evaluation, International Journal of Technology Management, Science and Public Policy, Technological Forecasting and Social Change, EASST Review, etc. and national ones.

Postal address: Consejo Superior de Investigaciones Científicas (CSIC), Unit on Comparative Policy and Politics, C/ Alfonso XII, n. 18, 28014 Madrid, Spain.

Email: lsanz@iesam.csic.es

Staman, Jan, DVM, L LM, graduated in veterinary medicine in 1975. He had till 1982 an appointment in small animal medicine in the Veterinary Faculty of the University of Utrecht. In 1988 he got his academic degree in Law. From 1982 till february 2002 Staman had several positions in the Dutch Ministry of Agriculture, Nature Management and Fisheries. His career started in the Veterinary Service and the Agricultural Education Inspection of the Ministry. From 1996 till 2002 he had an appointment in the Minister's Office in the Strategic Policy Division. Staman was during his career in the ministry involved with the ethical issues of Dutch agriculture, nature management and fisheries. In this field he had a leading and coordinating position. He was very much involved in the risks and ethics of animal biotechnology and the Dutch regulations in this field. During the last years he took the initiative for a Transatlantic Platform for Consumer Concerns and International Trade. He was also one of the initiators of the European Society for Agricultural and Food Ethics. During the last two years he made major efforts in the field of the food ethics in the Netherlands.

In February 2002 he got an appointment as Director of the Dutch institute for technology assessment, the Rathenau Institute. This institute covers major programs on Biomedical Technology, ICT, Animal Husbandry and Food. It also holds the Dutch Platform for Science, Technology and Ethics.

Postal address: Rathenau Institute, Koninginnegracht 56, 2514 AE Den Haag, Postbox 85525, 2508 CE Den Haag, The Netherlands.

Email: J.Staman@Rathenau.nl

Stephan, Susanne, studied economics at Cologne University for applied science (Rheinische Fachhochschule zu Köln) with core subjects on media and marketing and a thesis on "media cities in Germany". She has worked as assistant consultant at the consultancy company msc Multimedia Support Center GmbH in Bonn and BIA Europe, the European business incubator association in Malta; since 2002 she is a member of the Europäische Akademie GmbH in Bad Neuenahr-Ahrweiler with managing responsibilities for the project TAMI. Her research interests are the areas of private-public collaboration in Technology Assessment and the European project development.

Postal address: Europäische Akademie GmbH, Wilhelmstr. 56, Bad Neuenahr-Ahrweiler, Germany.

Email: susanne.stephan@dlr.de

Szapiro, Tomasz, Professor, is Professor in the Institute for Modern Civilization and Head of Division of Decision Analysis and Support, Warsaw School of Economics (WSE; SGH). Former Dean of WSE Graduate School (two terms). His degrees include: MSc in Physics from Warsaw University, Poland, PhD in Mathematics from Polish Academy of Sciences, habilitation and professor title in Economy. Adjunct Professor of International Business Studies at Carlson School of Business, University of Minneapolis, Minnesota, US. Research focused on Decision Analysis in Economy and Management. Publications in Poland and abroad (two books, monograph, over 120 articles and conference papers). Referee for international research journals. In 1989 visiting Professor and part-time lecturer in the School of Business, Carleton University and in the Department of Computer Science, University of Ottawa, Canada. Expert for UNIDO (trainings), for European Commission (project evaluator), Ministries of Finance, of Economy and of National Education.

Postal address: Institute of Modern Civilisation, ul. Koszykowa 80, 02-008 Warsaw, Poland.

Email: tszapiro@sgh.waw.pl, website: http://akson.sgh.waw.pl/~tszapiro

Steyaert, Stef, has a degree of Sociology (1989, University of Leuven). After a short period as a researcher at the University of Louvain, he joined the Flemish Foundation for Technology Assessment in May 1992. After three years of research in the field of biotechnology, he re-orientated his research work towards working conditions in public services, educational sector and financial sector. He published several books, studies and numerous articles on the topic of stress and working conditions. In 1999, he started working for nearly two years as a senior consultant in one of the major Belgian banks. In April 2001, he joined the Flemish Institution for Science and Technology Assessment where he became responsible for the methodological support of the conducted projects.

Postal address: viWTA-Samenleving en Technologie, Vlaams Parlement, 1011 Brussels, Belgium.

Email: stef.steyaert@vlaamsparlement.be

van Est, Rinie, Dr Ir., is both a physicist (MSc) and a political scientist (MA) and since 1997 has worked for the Rathenau Institute, the Dutch national technology assessment (TA) organisation. He deals with TA methodology, international projects and various TA topics. He has been involved in several European projects, like the EUROPTA project (1998–2000) on participatory TA methods, and the TAMI project (2002–2003). In 2002 he managed the project Towards a Societal Agenda for Food Genomics. Currently he is working on a project on the societal consequences of nanotechnology. Since 2000 he has been an Assistant Professor, lecturing at the Technical University Eindhoven on technology assessment and foresight. In 1999, he obtained his doctorate (PhD) from the University of Amsterdam. His thesis, titled "Winds of Change", analysed the interaction between policy and technical developments in the field of wind energy in Denmark and California.

Postal address: Rathenau Institute, Koninginnegracht 56, 2514 AE Den Haag, Postbox 85525, 2508 CE The Hague, The Netherlands.

Email: Q.vanEst@rathenau.nl

In der Reihe *Wissenschaftsethik und Technikfolgenbeurteilung* sind bisher erschienen: